William Brown

The Natural History of the Salmon

Ascertained by the Recent Experiments in the Artificial Spawning and

Hatching

William Brown

The Natural History of the Salmon
Ascertained by the Recent Experiments in the Artificial Spawning and Hatching

ISBN/EAN: 9783337025410

Printed in Europe, USA, Canada, Australia, Japan

Cover: Foto ©berggeist007 / pixelio.de

More available books at **www.hansebooks.com**

THE NATURAL HISTORY OF THE SALMON,

AS ASCERTAINED BY THE

RECENT EXPERIMENTS

IN THE

ARTIFICIAL SPAWNING AND HATCHING OF THE OVA AND REARING OF THE FRY, AT STORMONTFIELD, ON THE TAY.

BY

WILLIAM BROWN,
PERTH,

Secretary to the Literary and Antiquarian Society of Perth,
Author of "The Royal Palaces of Scotland," etc.

GLASGOW:
THOMAS MURRAY AND SON.
EDINBURGH: PATON AND RITCHIE.
LONDON: ARTHUR HALL, VIRTUE AND CO.
1862.

TO THE MEMBERS AND HONORARY MEMBERS OF

The West of Scotland Angling Club,

WHO FIRST INTRODUCED

THE GRAYLING TO SCOTLAND,

AND HAVE BEEN SUCCESSFUL IN ITS ARTIFICIAL
PROPAGATION IN THE CLYDE,

THE FOLLOWING ACCOUNT OF THE ARTIFICIAL
PROPAGATION AND REARING OF

THE SALMON

IS BY PERMISSION MOST RESPECTFULLY INSCRIBED
BY ONE OF THEIR NUMBER,

THE AUTHOR.

PREFACE.

THE writer of these pages had his attention first attracted to this subject by the experiments of Mr Shaw of Drumlanrig—from 1833 to 1838. Previous to that time, with very few exceptions the whole of those persons who were connected with the salmon fisheries believed that the parrs, which are so numerous in all salmon rivers and streams, were a species of fish (*sui generis*), and which never attained a larger size than seven or eight inches; the principal use of which fish was to afford sport to juvenile anglers, and to serve for bait in the capture of larger fish by more advanced "Waltons."

Being a believer in the general parr theory, the writer resolved to put Mr Shaw's statements to the test; and having, in the month of February, 1836, caught a dozen and a-half of parrs in the Tay, he kept them confined in a stream of running water, and by the month of May the whole of them had become smoults; but some had leaped out of their confinement, in their struggle to find their way to the sea, and were found dead upon the side of the pond. This having convinced him that what was called parr in salmon rivers, was, in fact, the young salmon ere it became a smoult, he entered with much earnestness into the artificial propagation scheme when it was started at Stormontfield, and was present and assisted at most of the manipulations.

The author would here record his best thanks to Robert Buist, Esq., superintendent of the Tay Fisheries (to whom belongs the merit of carrying on the experiment, for it required no small amount of intelligence and perseverance to unravel the seeming discrepancies of Shaw and Young in the natural history of the salmon up to the period of the smoult state), for the assistance which he has given the writer, while watching and taking part in the various experiments as they progressed. He would also acknowledge the services rendered by Andrew Buist, Esq., the present tacksman of Mr Ashworth's Galway Fishings, who, while the experiment was carrying on at Stormontfield, hatched the ova of salmon in a run of spring water at a steady temperature of 44 degrees F., in fifty days—spawned 18th January, hatched 9th March. Also by C. F. Walsh, Esq., Dundee (late of Perth), who was present at most of the operations which he so graphically described in the columns of the *Field* newspaper.

The descriptions given of the various processes are from notes taken on the spot by the writer; and as the notices of the Stormontfield experiment, which have been already published, are chiefly confined to paragraphs in newspapers, the reports before the British Association, etc., he sincerely trusts that the following Work will prove acceptable to the public, and at the same time tend to throw some farther light on the Natural History of the Salmon.

CROFT COTTAGE,
PERTH, *June*, 1862.

CONTENTS.

	Page
Introduction,	17
Commencement of the Stormontfield Experiment,	23
Hatching-Boxes and Pond,	29
First Artificial Spawning of the Salmon—Spawning of 1853,	36
Deposition of Ova,	88
Hatching of the Ova, 1854—Spawning of 1853,	40
First Migration of the Smoults, 1855—Hatching of 1854,	47
Hatching of 1855—Spawning of 1854,	48
Return as Grilse—Hatching of 1854,	49
Third Spawning, 1855,	54
First Salmon of the Spawning of 1853—Hatching of 1854,	54
1856 Second Migration of Fry Spawned in 1853—Hatching of 1854,	55
Exodus, 1857—Spawning of 1855—Hatching of 1856,	63
Fourth Spawning, 1857,	69
Hatching of 1858,	71
Exodus of 1859—Hatching of 1858,	75
Spawning of 1859,	77
Hatching of 1860,	78
Exodus of 1861,	80
Natural History of the Salmon as learned from the Stormontfield Experiment,	87
On the Advantages of the Artificial Propagation and Rearing of the Fry of Salmon until they reach the Migratory Period,	97
The Smoult or Young Salmon after Migration,	105
Return of Spawned Fish or Kelts as Clean Salmon,	106

CONTENTS.

	Page
Do Salmon fall off in Condition on entering the Fresh Water?	108
Kelts or Spawned Fish, ...	109
Attempt to raise the Salmon from the Ovum to the Grilse State Artificially, ...	116
APPENDIX, ...	123

ERRATA.

Page 26, 6th line from top, *for* "About 370" *read* "A number of."
Page 54, 10th line from top, *for* "Poynton" *read* "Egerton Hall."

SALMO SALAR.

DESCRIPTION.

HEAD, small in proportion to the size of the fish; maxillary bones, also the palate and tongue in the adults, set with sharp, stout teeth; gill-rays, never less than eight; opercle, posterior edge round, with a black spot in the centre; vertebræ, sixty; dorsal fins, two; the ventrals opposite to the middle of the first dorsal, the adipose dorsal opposite the anal. The number of fin-rays are as follows:—1st Dorsal, 12—Pectoral, 13—Ventral, 8—Anal, 10—Caudal, 19.

The adult male fish is easily distinguished by the lower maxillary bone and cartilage greatly protruding—this is very remarkable in spent or spawned fish. The tail of the full-grown salmon is straight across, while in the grilse and young salmon it is forked.

THE

NATURAL HISTORY OF THE SALMON.

INTRODUCTION.

THE natural history of the salmon, previous to the investigations and experiments of Messrs Shaw and Young, was unknown. The general opinion was that the salmon ascended the rivers during the summer and autumn months, and spawned in November and December; the fry having been hatched in March and April, descended the rivers as smoults in May and June following. Whether our forefathers were of the same opinion, we have no means of knowing; but we suspect, from the wise and salutary laws which they made from time to time for the protection of this fish, that they knew more about its habits than was known at the commencement of this century. The Act of James I.,* 1424, c. 35, continued in force up to 1828, when it was repealed by the Act 9 Geo. IV., c. 39—commonly known as "Home Drummond's Act," which be-

* See Appendix.

came law in 1828. For four centuries this old Act of James I. regulated the close time in all the rivers of Scotland, except the Tweed and the Solway Frith and its tributaries; and which proved very efficient for the protection of the spawning fish, as the close time commenced on the 15th of August, o.s., instead of—as by Home Drummond's Act—on the 14th of September. This additional month of open time allowed by Home Drummond's Act to all methods of capturing salmon, both by net and coble, and fixed engines on the sea-coast, has gone far to accomplish the destruction of the fish in all the Scottish rivers. Close time on the sea-coast, to be equal to the same length of time on the rivers, should commence a fortnight sooner; as a fish, caught on the coast, many miles away from the mouth of the river which it was trying to reach in order to ascend for the purpose of spawning, would take at least that time to reach the net and coble, on the river; and hence the number of fish full of spawn, and nearly ready for spawning, that are caught at present from the 15th of August (the commencement of the old time) until the 14th of September (the commencement of the present close time). So, if we are to judge of the knowledge of our ancestors by the Act of James I., we must come to the conclusion that they knew more of the habits of the salmon than was known forty years ago; besides, for the four centuries in which this old Act was in force, no stake or bag nets were to be found on our coasts. Other Acts were, from time to time, made for facilitating the ascent of the spawning fish and the descent of the smoults; weirs, cruives, etc., were particularly attended to, so that the fish might have clear way. "Our Sovereign Lord and the Three Estates of Parliament" seem to have looked to the protection of salmon in a national point of view—as food for the common people; for far greater protection appears to have been given to the spawning fish and the fry, and the penalties were more severe for infringing the various Acts than at present. Hence we might conclude, from the attention given to these subjects, that more was known of the natural history of the salmon by our ancestors than by us.

It is really astonishing to mark the prejudice and gross ignorance of the natural history of this fish, displayed by practical fishers at the present time—some of whom have been for forty years connected with salmon fishings—as exhibited in, their evidence before the Select Committee of the House of Lords in 1860.* We look in vain for anything like the same want of knowledge amongst the proprietors and rearers of stock on the land. They are all alive to ascertain the true history and character of the stock they deal in, and also the best methods of increasing its numbers and value as a marketable commodity. A landowner, who has purchased a property on which there is a small stock of game, when he wants to increase it, does not shoot down what remains, but re-stocks it from other properties where the game is more plentiful, and carefully destroying all the vermin that would prey upon it. A knowledge of the history and habits of the denizens of the water is certainly more difficult to acquire than those of the land; but with regard to the salmon, while in the fresh water at least, this has been accomplished by those who have witnessed the conducting of the Stormontfield operations; and all that is wanting for those who

* Appointed to inquire whether, having regard to the right of property of the Crown and individuals in salmon fishings on the sea-coasts and in the rivers and estuaries of Scotland, it is just and expedient that any, and what, legislation should take place for the regulation of such fishings; so far as regards the use or prohibition of bag nets, cruives, and other fixed nets and engines, and so far as close times or otherwise.

would learn something more of the history and habits of this noble fish, is to lay aside their previously acquired notions and prejudices, and to believe the *facts* which have been brought to light by those most interesting and successful experiments.

NATURAL HISTORY OF THE SALMON.

STORMONTFIELD EXPERIMENT.

In the summer of 1853, the late Dr Esdaile of Perth published a letter, addressed to the salmon fishing proprietors of the Tay, on the artificial propagation of salmon, the result of which was that a meeting of that body was held on the 19th July. Dr Esdaile was present, and stated that his attention had been called to the subject by what had been written by Mr Thomas Ashworth of Poynton, Cheshire, who, along with his brother, Mr Edmund Ashworth, had purchased the Galway salmon fishery in the Court for the sale of encumbered estates, Ireland, as a several fishery extending from Lough Corrib to the sea. Dr Esdaile then introduced Mr Ashworth to the meeting,* who said he had come to this part of the country to

* From the *Perthshire Courier*.

gather all the information he could obtain, having seen the letter of Dr Esdaile, addressed to salmon fishing proprietors in the Tay, on the "Artificial Propagation of the Salmon;" and it was at Dr Esdaile's request that he had consented to stay in Perth another day to attend the present meeting. He (Mr Ashworth) had entertained the opinion for a long time that it would be as easy artificially to propagate salmon in our rivers as it was to raise silk-worms on mulberry leaves, though the former were under water and the latter in the open air. He said that it was an established fact, that salmon and other fish may be propagated artificially in ponds in millions, at a small cost, and thus be protected from their natural enemies for the first year of their existence, after which they will be much more capable of protecting themselves than can be the case in the early stages of their existence. His brother and he have at the present time about 20,000 young salmon in ponds, thus produced, which are daily fed with suitable food. In the course of last year, Mr Thomas Garnet and Mr Peel of Lancashire informed his brother and himself that Robert Ramsbottom, a fishing-tackle maker at Clitheroe, had, about two years ago, succeeded in artificially filling a small pond with young salmon. The plan adopted by him was a very simple one. He, with great care, lifted the salmon from the river into a tub of water, and extracted their spawn, which was deposited in boxes filled with water and gravel; they were placed in a running stream, and by this means had produced the young salmon. His brother, Mr Edmund Ashworth, and himself having purchased "the Galway salmon fishery" in Ireland, they determined to try an experiment there for the artificial propagation of salmon, under the superintendence of Mr Ramsbottom. A suitable place having been fixed upon at Outerard, operations were commenced between the 20th December and the 1st of January last, which was about a month too late, yet boxes were prepared in which the spawn of the salmon was deposited. These boxes were about 18 inches broad and deep, and 6 feet in length, with a zinc grating in the sluice at either end. There were twenty boxes in all, which were filled with gravel or small stones to the depth of six inches. To procure the ova and milt of the female and male salmon, the fish were taken by small nets on the spawn fords at night, and instantly, and without injury, put into a tub one-fourth full of water. The female fish was first turned on her back, one man holding the tail, another running his hands down each side from the head, and pressing lightly with his thumbs, the ova were readily discharged into the tub; a similar course readily discharged the milt. About 370 salmon were treated in the above manner, and again returned to the river. Mr Ashworth then explained how the ova and milt were mixed in the tub, and the ova then taken out of it with a cup and deposited in the boxes, when they were covered with additional gravel. Mr Ramsbottom supposed that there was put in the boxes spawn sufficient for 40,000; however that might be, there are at present about 20,000 young salmon alive and thriving in these ponds, and from two inches to three inches in length. Mr Ashworth then, in answer to several questions put by Mr Graham of Redgorton, stated that the fine zinc gratings were used to prevent trout and insects from getting into the ponds, as they are very destructive to the salmon fry. He then detailed, at some length, the mode adopted in France for the artificial propagation of salmon, carp, trout, and other species of fish. The boxes there used were 3 feet in depth, 18 inches in width, and 50 yards in length, and without bottoms, but filled about 6 inches in depth with clean gravel. In answer to other questions, Mr Ashworth said the ponds were about 20 yards in length, and 12 to 13 yards in breadth, and that it was intended to keep in the young salmon for ten months, when they will have grown to about four inches in length. They would then be able to take care of themselves on their way to the sea. He stated also, that it was indispensable the young salmon should be fed daily with chopped flesh meat. In France they were fed on frog's flesh, pounded quite small in a mortar. The current of water running through the boxes must be pure and free from mud, and great care required to be taken during the periods of hatching, if the rivers were flooded by heavy rains, to divert the muddy water from the boxes. It took about 100 days until the spawn gave indication of life. Mr Ashworth then observed that a great deal had yet to be discovered in the artificial propagation and feeding of salmon. They knew comparatively little of the habits of salmon, and in order that a greater amount of knowledge might be obtained, he had recommended to the Commissioners of Fisheries in Ireland to take from the ponds a portion of the fish propagated in the way he had mentioned, and immerse them annually in the sea for a period of three months, and to be again deposited in the ponds for other nine months, to be repeated for several years. The Commissioners had taken about a dozen of these young salmon from the ponds, and had had them many weeks in the Dublin Exhibition, where they were kept in a model of a weir, with a salmon ladder in it, the model being supplied by a pipe with a constant run of water. These little creatures showed their agility by mounting the ladder, and so passing over the weir, to the amusement of the bystanders. He was informed they were alive, thriving and perfectly healthy in this small run of pure water; they were fed on chopped meat every day. It was only in this way a more accurate history of the ages and habits of the salmon species might be written. The expense of this plan of artificial propagation he did not estimate to exceed a pound a thousand, which was at the rate of one farthing for each salmon.

At the conclusion of Mr Ashworth's lucid remarks, Sir Patrick M. Thriepland moved that

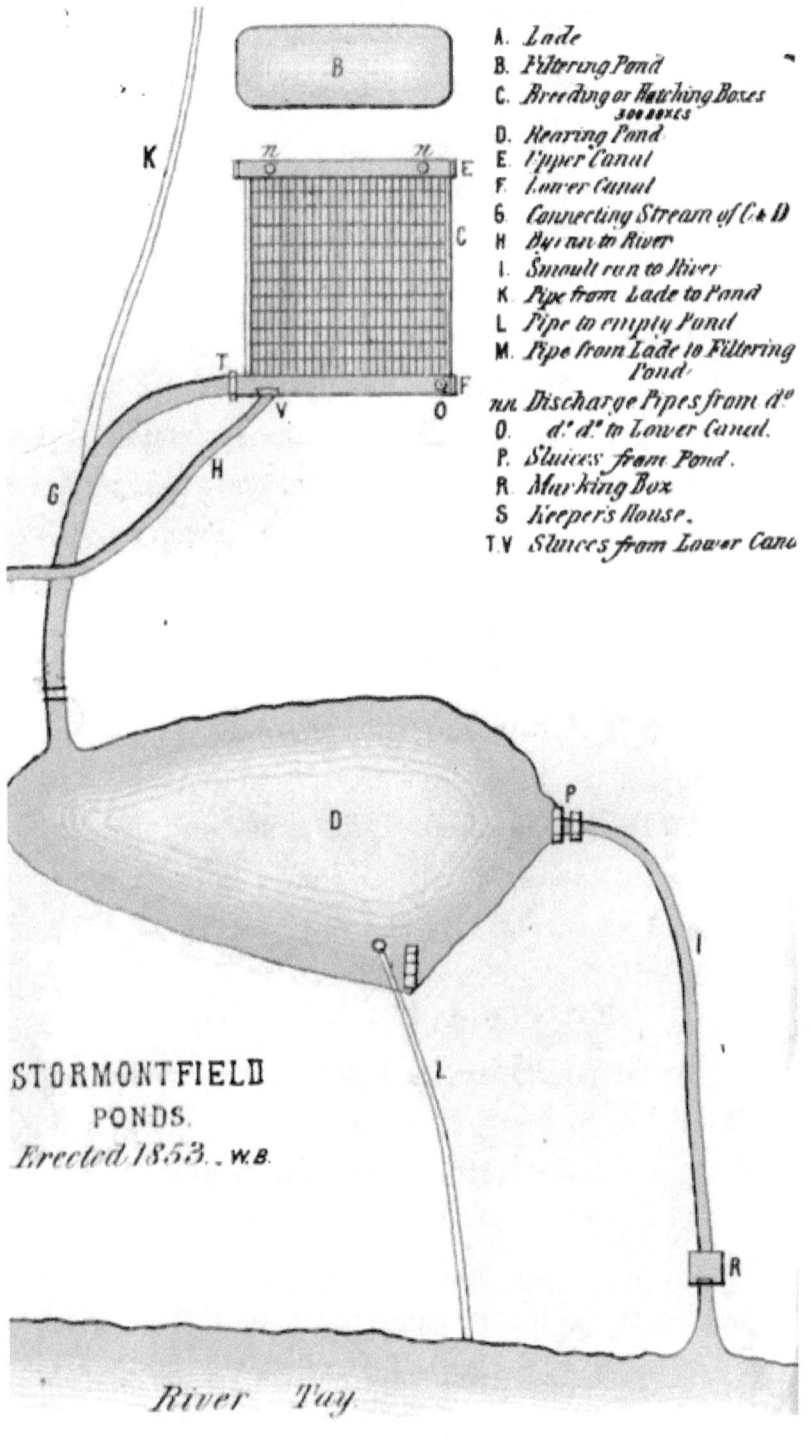

the thanks of the meeting be tendered to that gentleman for the important information which he had communicated.

Colonel Belshes seconded the motion, which, being unanimously agreed to, the Chairman conveyed the sense of the meeting to Mr Ashworth.

At another meeting held on the 21st October following, it was resolved to make ponds and rearing-boxes, and £500 was voted for that purpose. A Committee was also appointed to superintend the work.

Immediately after this meeting operations were commenced, and by the 23d of November the boxes were in readiness to receive the ova. The accompanying plan will be better understood, and answer the purpose more fully than any description we could give of the ponds and hatching-boxes.

HATCHING-BOXES AND POND.

This far-famed salmon-hatching and fry-rearing establishment is situated on the river Tay, about five miles above Perth. Lord Mansfield is the proprietor of the ground. The hatching-boxes are 16 feet above the summer level of the river, and the water is taken from the lade that supplies the Stormontfield bleachfield; this lade comes off the Tay about a mile farther up the river than the ponds. As will be seen in the plan, a pipe (the mouth of which is covered with perforated zinc) is taken from the lade to the filtering pond, the water from which, after being filtered, rises into the upper canal, and flows through the hatching-boxes, which empty themselves into the lower canal, the water from thence escaping into the rearing pond by means of a small artificial stream plentifully supplied with large gravel as shelter to the young fish. A current is kept constantly running through the pond, at the east end of which there is an outlet having a sluice; and just before the water enters the river there is another sluice, for intercepting the descending smoults, for the purpose of marking. It will be seen by the plan that there are 300 boxes in all, laid out into 25 parallel rows, with a walk or path between each, twelve boxes being in in the row, each box 6 feet long by 1 foot 6 inches broad and 1 foot deep. The fall from the upper to the lower end of each box is 2 inches, and 2 feet in all, so as to allow the water to flow freely through them; but experience has proved that the fall is not sufficient, and that the quantity of water discharged from the feeding pipe is too small. A larger run of water and a greater fall would scour the gravel better, and keep it freer of algæ and impurities of all kinds. The boxes are filled to within an inch or two of the top, first with a layer of fine gravel, next with one of coarser gravel, and lastly with stones as large as road metal. These deposits, previous to being put into the boxes, should be washed clean, and allowed to lie exposed to the action of the sun and air, in order to rid them of the larvæ of insects, which destroy more ova than all other enemies put together.

When the young fry seek change of quarters, they drop down to the lower canal, where they are regularly fed. At this time the canal is swarming with them—there are more fish there, were they allowed to reach to the full-grown state, than would pay the rental of the Tay for one year. All the fry do not leave the boxes at the same time; many of them linger on through the summer. As will be seen by looking at the plan, there is a connecting stream with the pond, which the keeper shuts off with a sluice; this he keeps shut until all the smoults are out of the pond, which is emptied and cleaned ere this sluice is opened and the young fry admitted, and that is generally done in the end of May. When the fry are liberated from the lower canal, they spread themselves over the pond, and into whatever part of it the keeper throws the food, they may be seen rising to it in great numbers. This pond is the most faulty part of the whole plan; it is only 223 feet long by 112 feet wide at its broadest part, and is far too small to contain and nourish 300,000 fry, which the boxes can hatch; it would require a pond of four times the size to rear the produce of the boxes; besides, another pond is necessary for enabling the hatching to be carried on every year. There is a slight current in the pond; but we think not so great as it should be, to supply abundant fresh water for so many fish. At the lower end of the pond there is a pipe, for the purpose of emptying it, connected with a drain to the river. When artificial propagation is more in favour than it is at present with our salmon proprietors, all these evils will be remedied. We have said that the gravel for the hatching-boxes should be well cleaned, dried, and exposed to the action of the sun and air, in order to rid it of the larvæ of insects, the greatest enemy the ova have, which statement we shall illustrate thus:—

In the spring of the year 1854, Mr Buist, the conservator of the Tay, obtained some ova, nearly ready for hatching, from a ford on the river, and placed a dozen of the grubs of the May-fly (*ephemera*), taken from the same bed, along with them in a vessel, which was supplied with water by a syphon of thread. In a few days the grubs had devoured one of the eggs, and in a few days more the whole were devoured; but, previous to that time, two or three of the grubs left their covering, and came forth as the May-fly. We watched them carefully while in the act of feeding, and found five or six of the grubs firmly fixed to an ovum, which they never left until totally eaten up. These animals are not the scavengers of the river, for, in this instance, the ova were alive. Again, in a small but complete artificial rearing apparatus, which we have had in operation for many years, and which is supplied with filtered water, we deposited in two boxes, on the 26th November, 1859, a quantity of salmon ova, fecundated by the milt of a male salmon, and on the 30th of the same month a small quantity of sea-trout ova, fecundated by the milt of a male salmon also. The progress made by both was very satisfactory;

the temperature of the water was 40° when the ova were deposited—never falling below 36°, and by the 1st of March the eye and round form of the fish could easily be detected in both kinds, by the naked eye, and an ovum, when put in the hollow of the hand, would turn itself round. Peter Marshall, the keeper at the Stormontfield pond, who was in the habit of examining them regularly, stated that they were about a fortnight earlier than the ova at that place, which had been deposited at the same date. But about this time, on account of a deficient supply of filtered water, a quantity of unfiltered water was allowed to enter the pipe; this water contained a large amount of the larvæ and grubs of insects, particularly of a small black water beetle, and by the end of April all the ova were devoured. Their method of procedure was as follows:—the grub fastened on a live ovum, and pierced a hole in the shell, the colour instantly changing from a salmon colour to opaque white: the egg was devoured at leisure afterwards. The Messrs Ashworth, proprietors of the Galway fishings, experimented on the May-fly, and their report is, "that the larvæ of the May-fly are known to be most destructive;" in proof of this being the case, they say—"that one year we deposited 70,000 salmon ova in a small, pure stream, adjoining to a plantation of fir-trees, and these ova we found to be entirely destroyed by the larvæ of the May-fly, which, in their matured state, become the favourite food of smoults or young salmon. It is evident that insects, fish, and fowls destroy by far the largest quantity of the salmon species, and they cannot be removed as you would destroy vermin in a game preserve; it is therefore reasonable to suppose that the parent salmon is induced instinctively to surmount the greatest difficulties, and to incur the risk of its own life by ascending to the smallest streams, where alone its offspring can be most securely deposited and reared beyond the reach of its numerous natural enemies. We know that the natural enemies of the salmon cannot be destroyed, as they exist both in rivers and in the sea, consequently there is left but one certain mode of increasing the quantity of salmon, and that is by artificial means—by collecting the spawn, and placing it beyond the reach of its enemies for hatching and protection for the first year of its existence; and this may be done, in vast quantities, at a small cost, and without injury to the parent fish."

From these experiments, we may see the necessity of having the gravel free from the eggs of grubs and larvæ of insects, and likewise learn the great destruction of ova that takes place by their means in the beds of all our rivers and streams. It is true that insects, in all their stages, become, in their turn, the chief food of the young fish, and that no river could feed or support fish, if these were absent; but as the fry of the salmon, at least, require no external nourishment (as we shall show afterwards) for the period of six weeks after hatching, the insects can be of no use to them for food, and, in artificial rearing, food is supplied them by their keeper. There is little doubt but that the most of our salmon rivers (if well protected) would be sufficiently stocked with fish, if they were not, in every instance, over-fished by nets of all kinds; but when that is the case, artificial rearing, if generally adopted on a scale sufficiently large for the size of the river, would rescue enough of the ova from their enemies to meet this extra fishing.

ARTIFICIAL SPAWNING OF THE SALMON—SPAWNING OF 1853.

On the 23d of November, 1853, this operation

was commenced at a ford on the Tay, near Almond-mouth, by Mr Ramsbottom of Clitheroe, who is so well known as a successful manipulator. We were present, but will let Mr Ramsbottom explain his own method, which we take from his pamphlet on Artificial Breeding at Outerard:—
"So soon as a pair of suitable fish were captured, the ova of the female were immediately discharged into a tub, one-fourth full of water, by a gentle pressure of the hands, from the thorax downwards. The milt of the male was ejected in a similar manner, and the contents of the tub stirred with the hand. After the lapse of a minute the water was poured off, with the exception of sufficient to keep the ova submerged—this must always be attended to, even when the ova or milt is flowing from the fish—and fresh water supplied in its place. This also was poured off, and fresh substituted, previous to removing the impregnated spawn to the boxes prepared for its reception." We observed in this, the first manipulation, and in all the others afterwards, that a very small quantity of the milt was sufficient to impregnate the ova of a large salmon, and that always a few of the ova, after receiving the milt, turned white—these were injured, and

would prove addled. We also noticed that the salmon colour of the ova was heightened when the milt came in contact with it. Round tin pans, with as much water in them as covered the ova, were used to carry them to the hatching-boxes. On the 23d November, 1853, the first stocking of the boxes commenced, and by the 23d of December 300,000 ova, in very fine condition, were deposited in the 300 boxes. The spawned fish were returned to the river, and went away after the operation quite lively.

DEPOSITION OF OVA.

The ova were sown into the boxes, which had been previously prepared, and chiefly at the top, by means of the lid of the pan, as the stream which flows through carried them to the lowest part, but it is necessary to assist the current when it is not strong enough, by pouring a few pails-full of water in at the top of the box, or agitating the water with the lid so as to insure the proper and equal spreading of the ova; also, care should be taken not to sow it too thick, as ova, it has been found, thrive best when they do not touch one another in the spawn bed. The season of 1853–4 turning out very severe, the hatching was retarded, the temperature of the water flowing through the boxes during the month of January, February, and March ranging between 33 and 42 degrees Fah.

After Mr Ramsbottom had seen the ova deposited in the boxes, Mr Buist selected to attend and watch over the young fry Mr Peter Marshall, the present keeper, who has shown all along an aptitude for the situation; and we know he has given much satisfaction to his employers, and to all naturalists and strangers who have visited the place to witness the experiment. We question if there is any person in this country more thoroughly acquainted with the artificial rearing process than Peter of the Pools.

Many theories were prevalent at this time—the public prints were full of them; but there was one that Mr Buist was determined to examine, namely, whether ova taken from the female fish, without having received the milt, would hatch, some writers affirming that the ova were impregnated before they left the fish. Two boxes were filled with ova taken from a salmon without having the milt applied, and other two were filled from the same fish, but manipulated upon with the milt of the male; and it was found that not one of the former hatched, while all the others succeeded. What was singular in this experiment was, the barren ova only became a little paler, but never turned white or opaque like ova that had been impregnated and died afterwards. We examined many of these ova in the months of April and May (when all the boxes, except the two reserved for the experiment, were swarming with fry), and there was not a chick visible in any of them; therefore we have not the least doubt that the ova of the salmon at least are not impregnated until they leave the fish. The number of ova in a female salmon is roughly reckoned by the weight of the fish, as it has been ascertained by counting the ova, that for every pound weight she will shed about 1000 ova. A salmon of 10 lbs. will therefore contain about 10,000 ova. The keeper informs us that the best period for transporting salmon ova to any distance is immediately after impregnation, or within a week or two of the time of hatching.

HATCHING OF THE OVA, 1854.

On the 31st of March, 1854, the first ovum was observed to have hatched, which was 128 days from the deposition of the first and 98 days from the deposition of the last of the ova. A high or low temperature of the water will accelerate or retard the hatching; ova have been hatched by us in 60 days, in a constant temperature of 44 degrees, but in the rivers in this latitude from 100 to 140 is the time, according to the season.

We were furnished with a few ova, and, by keeping up a supply of pure water we were gratified by observing the little creature bursting the shell. The fish lies in the shell, coiled round in the form of a bow, and the greatest strain being at the back, it is the first part that is freed; and, after a few struggles, the shell is entirely thrown off with a jerk. The appearance of the fish at this stage of its being is very interesting; what is to be the future fish is a mere line, the head and eyes large, the latter very prominent. Along the belly of the fish, from the gills, is suspended a bag—of large dimensions in proportion to the size of the fish. This bag contains a yolk which nourishes the fish for six weeks, after which they must be fed. For a few days after hatching, the two dorsal fins are apparently joined, and the two pectoral are very large in proportion to the rest of the animal. The little creature, not requiring to seek its food, moves very little, and when it does, swims mostly on its side, owing to the large size of the bag; this gradually becomes absorbed, and in a short time the fins get separated, and the fry assumes the general aspect of a fish. In its first stage it is translucent, but in a short period it takes on the parr colour, and the transverse bars can be easily seen, and the tail begins to get much forked. At the bag stage of their existence they are very easily injured; a displaced stone in the gravel amongst which they are lying coming against them destroys them; and although they are no longer the prey of insects, all kinds of fish and fowl are their enemies, and great must be their destruction in rivers where their enemies are numerous. As we have previously stated, in about six weeks the bag is absorbed, and the fish is a fingerling or parr, from one inch and a half to two inches long. The pond in connection with the boxes was ready to receive the fry when the fish had reached this stage, but was found far too small to receive 300,000 fry. The length of the pond is only 223 feet by 112 feet at its broadest part, but from the great attention paid to the fry by the keeper, and the regularity with which they were fed with boiled ox and sheep liver ground small, almost no deaths occurred, and by the time the fry were a year old, by far the greater part had grown as large as the parrs found in the river; and well they might, for they fed greedily, and it was an interesting sight to witness them getting their food.

We were at a loss to account for the difference in size which was discoverable amongst the fry at this period, for all were within a few weeks of the same age—the keeper thought that the small ones were kept by the larger ones from getting as much food as they required. The same difference in size at this time is found to be the rule also in the river, and those that we have reared ourselves, though equally and liberally fed, still showed the same difference. It was thought that the small fry might be from the ova of grilse, but this theory was put to the test afterwards, by keeping the fry of salmon and grilse distinct, and, amongst both, the same disparity in size appeared. Mr Shaw of Drumlanrig, the first who had experimented successfully on the artificial hatching and rearing of salmon fry, stated that the fish in the pond would not become smoults until another year; while Mr Andrew Young of Invershin, who had carried out with equal success similar experiments, asserted that the whole would become smoults, or, as parrs, migrate to the sea that season; and there was no little speculation created amongst those who took an interest in these matters to learn the result. Articles on the subject appeared from time to time in the *Perthshire Courier*, the *North British Daily Mail*, the *Scotsman*, the *Field*, and even the *Times* took up the subject.

The spring of 1855 was very severe, for the ice was not off the pond from the middle of January till the 8th of March, and when the ice was at its thickest it was 10 inches thick, and chaffers, containing fire, had to be used at the outgoing sluice, so as to insure a constant run of water through the pond. Large holes were kept open in the ice, and the keeper, during the severest of the frost, had to visit them two or three times during the night to keep them open. At these holes the fish were regularly fed, and they appeared to feed greedily. This severe weather had an effect in retarding the development of the fry. On the 2d of May, a meeting of the Committee appointed by the Tay fishing proprietors was held at the pond, when Mr

Wilson of Woodville, Edinburgh, Mr Shaw, Mr Ramsbottom, Dr Esdaile, Mr Buist, Mr Walsh, and some other gentlemen who took an interest in the matter, were present. Lord Mansfield was also present, and took a prominent part in the proceedings, and entered very fully into the various discussions that followed on the specimens exhibited.

When the Committee met, a tub, containing a dozen or two of fry, which the fishermen had got in the river, was shown, for comparison with those in the pond; these fish were pronounced smoults by the fishermen, and Mr Shaw, who was provided with a small net, cast it into the pond, and brought out a number of fry of all sizes, which were put into another vessel.

The fry, on being compared, were very much alike in size and colour, only the pond ones were fatter, and Mr Shaw pronounced them (both river and pond) all one-year parr. In the meanwhile, Dr Esdaile caught a larger specimen with the rod in the Tay, which Mr Shaw at once said was a smoult on its way to the sea. Here was now the means of comparing the pond fish with the acknowledged real smoult, and settling the question. This last caught specimen was certainly larger than any that had as yet been taken out of the pond, and the smoult scales fuller grown—we have no doubt but it was an early smoult from some of the higher tributaries—but the parr marks were rather more visible than on the pond fish. The back was of a dark blue colour, and the pectoral fins were getting black, while the backs of the pond fish and those in the tub were yellowish brown. It was suggested that this brown colour of the back might be owing to the fish assuming the colour of the tub, and Mr Shaw showed that it was so, by putting some of the fish into dark and light-coloured vessels, and it was seen that the fish in a very short time became dark or light according to the colour of the vessel; but the pectoral fins were still dark in the large specimen, while the others were yellow. Messrs. Shaw and Wilson, after an interesting discussion, came to the conclusion, which the Committee adopted, that the fry in the pond were still parrs, and would not migrate until next spring. We expressed our conviction that the Committee would find that they were mistaken, judging from previous experiments which we had made, and knowing well the mind of our friend Mr Young on this subject, we asked Mr Shaw, subsequently, how it was that so many of the pond fish had got on the smoult scales, if not ready to go to sea? Mr Shaw said that all parrs at this season put on these scales but threw them off again. We replied that this was not in accordance with our experience. We again asked Mr Shaw if none of the pond fish would seek to go to sea? He said some might go, but not many; although he had seen them go the first year. The Committee then left, and the keeper was ordered to retain the fish in the pond; but some of the tacksmen of the salmon fishings on the Tay, who were present, were so convinced that the fish were smoults, that they threatened not to pay their rents if the smoults were not allowed to go.

FIRST MIGRATION OF THE SMOULTS, 1855.

On the 19th of May, Mr Buist becoming convinced that the fry had become smoults, *i.e.*, had taken on the silvery scales, caused a great many to be marked by cutting off the dead or second dorsal fin, and turning them into the river. The sluice was drawn, and the fish were allowed to depart; but, contrary to expectation, almost none showed any inclination to leave until the 24th of May, when the exodus began, and a shoal came down to the marking-place, when the keeper was engaged at the time in marking, amounting, according to his calculation, to something like 5000, and so full of fish was the marking-trough, that he had to desist and let them escape into the river without being marked. The mark this year was cutting off the second dorsal fin. By the 3d of June the keeper had marked 1300, and it was quite evident that many of the fry would not seek to migrate this season, as they had not taken on the smoult dress. An article in the *Scotsman* newspaper, from an English correspondent, contained a theory, that the remaining parrs would be all males, as it would account for the number of large male parrs found in the river in the months of October, November, and December. We dissected a number as they came to hand, and found that in the remaining fish the sexes were about equally divided, so this theory was set at rest; and although some of the Committee, and almost all the fishermen, were for turning all the fry out of the pond into the river, it was decided to allow them to remain for another year with free egress to the river.

The hatching of 1855 proving a failure—not more than a few thousand fry having been produced, owing to unskilful manipulation in the spawning of the fish, (which was not entrusted to Peter,) and the long severe cold of the season—they were therefore put into the river; and it was agreed in future, as there was but one rearing pond, to hatch only every second year, as it had now been ascertained that only one-half of the former hatching would leave the pond the first year, and to turn in among the full-grown parr, the new hatching would be certain destruction to it.

RETURN AS GRILSE.

On the 7th of July, 1855, the first marked grilse was caught, returning from the sea, at a fishing-station near the mouth of the river Earn—a tributary joining the Tay, a little below Perth—which was about six weeks after the first exodus from the pond, on the 24th of May, This grilse weighed 3 lbs., certainly a great growth in so short a time—as the weight of a smoult before it reaches the tidal wave is from one to two ounces. The capture of this fish created some sensation amongst those who were watching the experiment. By the end of July, Mr Buist, the superintendent, stated that twenty-two grilse having the mark had been caught, and carefully examined by him, besides as many reported but not brought to him; and, but for the aversion shown by the fishermen to the experiment, he felt certain many more would have been reported. In order to induce the fishermen to bring the marked fish to Mr Buist, he caused the following notice to be circulated amongst them:—"A number of the smoults from Stormontfield breeding-pond having been marked by cutting the dead fin in a certain way, notice is hereby given, that 2s per lb. will be paid for each of the first five grilses that may be brought, so marked, to Mr Buist's office. The above price will be paid on its being satisfactorily ascertained that the mark corresponds with that made on the smoults.—Perth, 12th July, 1855." This was a tempting reward; and, as was expected, would induce some of the men to counterfeit the mark, and such was the case. A fisherman caught a grilse, and having cut off the dead fin, found that the raw wound would be easily detected; and, in order to disguise the part, he took a portion of the skin from off another grilse and put it over the place in the best manner he could; but Mr Buist instantly detected the cheat. In fact, it would be impossible to imitate the natural healing of the wound, as new skin and new scales grow over the part. The rapid growth of the grilse may be inferred from the weight of the first marked one which was taken, on the 7th July, being 3 lbs.; their weight increasing progressively to 5, 5½, 7 and 8 lbs., while one captured on the 31st July weighed no less than 9½ lbs. By almost all those in any way acquainted with the progress of the experiment, this was considered a very satisfactory elucidation of the history of the smoult up to the grilse state. But others, again, affirmed that the cutting of the fin might have been done by seals while they were in the sea. Seals will undoubtedly cut off the dead fin of a salmon or grilse; although a piece more of the fish is likely to be taken, and the fish mutilated. But it is a very wide stretch of the imagination to believe that 44 grilse captured in one season should have been deprived of this fin by these animals, and the fish be otherwise uninjured. It was, however, considered very desirable, in order to meet all objections, that some foreign substance should be inserted into the smoults next year, so that on their return they should be at once recognised. In the *Field* newspaper of the 12th of April, 1856, Mr

Buist makes the following statement:—"Believing that your readers would like to know the dates and the number of smoults marked as they left the pond, I append the necessary particulars. The numbers marked were about one in every 100 that left, so that about 120,000 departed, and more than double their number remained.

"Smoults marked on going out of the pond—

May 19,	41	May 31,	12
" 23,	25	June 1,	204
" 24,	48	" 2,	6
" 25,	58	" 3,	3
" 26,	24	" 4,	6
" 27,	114	" 5,	130
" 28,	56	" 6,	150
" 29,	374	" 7,	4
" 30,	37		
			515
	777		777
			1292

"The marked grilses were taken and reported at the following rates, the greater part having been produced and carefully inspected:—*

July 7,...1 grilse 3 lbs.	July 30,...1 grilse 7½ lbs.
" 20,...1 " 5½ lbs.	" 31,...1 " 9½ lbs.
" 24,...1 " 5 lbs.	Aug. 4,...1 " 7½ lbs.
	" 4,...1 " 8 lbs."

About 22 were inspected by Mr Buist, and 20 reported, but not seen, by him.

Allowing for what were taken with the marks and not reported, we may estimate at least forty, so that, taking them as one for every 100, would give 4000 grilses "having been added to the stock in the river from the breeding ponds." We are unable to give the date when taken, and weight of the other grilse seen by Mr Buist, as he has mislaid his jotting. The fish that remained in the pond by the end of the year, although healthy, did not increase much in size; but many of the male parrs were full of milt, and great numbers of the fry were still very small— not being much above three or four inches in length. To account for the great difference in size, it was thought that the large fish prevented the weaker and smaller from getting their regular share of the food, but, at the exodus of the smoults, we observed, and our observation has since then been confirmed, that both large and small alike take on the smoult dress when the migration time arrives, and those parrs that remain are composed in the same ratio of large and small. It was then conjectured that the small were the produce of two grilses, or of a grilse and salmon; but, as we proceed, we shall see that this conjecture was equally at fault.

THIRD SPAWNING, 1855.

On the 22d of November, 1855, the third spawning of the fish commenced, and continued until the 16th December, at the same place—the junction of the river Almond with the Tay—and, as the previous year had been a failure, Mr Ramsbottom was present to officiate. The total number of boxes filled on this occasion was 183, each box holding about 1000 ova. Mr T. Ashworth of Poynton was present at the operation. One of the male salmon taken on this occasion had been caught and marked by Mr Ramsbottom, in 1853, and had returned again to *spawn at the same place*. The first of the fry of this hatching came to life on the 3d of April, 1856—the rest from eight to fourteen days afterwards.

FIRST SALMON OF THE SPAWNING OF 1853.

On Saturday, the 1st March, 1856, the first salmon belonging to the hatching of 1854 was reported by Mr Young, fisherman, as caught by him. This fish weighed $10\frac{1}{2}$ lbs. We have no doubt, as we shall hereafter endeavour to prove, that this spring fish had not ascended the river the previous year as a grilse, but had remained all the winter in the sea. On the 8th of May another marked salmon, weighing 9 lbs., was caught at Kinfauns station; but the fishermen all along not only showed a want of interest in the progress of the experiment, but, in many instances, were decidedly hostile to it.

1856 SECOND MIGRATION OF FRY SPAWNED 1853 TO 1854.

Previous to the migration of the smoults this year, some of those gentlemen who felt an interest in the experiment thought, along with Mr Buist, that, in order to identify the fish on its return from the sea, it was desirable that some substance should be attached to the smoults, so that they should be recognised beyond all doubt on their return. Whatever substance was to be used, it required to be something that salt water would not destroy. Many things for marking and methods of attaching them were thought of, but our friend, Alexander Harvey, Esq., Glasgow, sent 300 silver rings, which we had great hopes would answer the purpose. These rings were pointed at one end, and had a loop-eye at the other into which the point end was inserted after being attached to the fish through the fleshy part of the tail—the pointed end being doubled back. The other mark this year was cutting off a portion of the lower angle of the tail. On the 28th of April, which was a month earlier than last year, the first shoal of smoults went off. They were generally larger, fuller grown, and finer shaped than last year. We took many smoults from the river for the purpose of comparison, and in nearly every instance the pond smoult was superior in every point of view. There were no very small parrs in the pond at this time as there were last year. Previously to leaving they congregate and swim in a restless manner through the pond; then a few of them visit the place of exit, and again return to the shoal, which shortly takes its departure; and it is a fact well known to all fishermen that those in the river descend to the sea in shoals. The keeper having reported to Mr Buist that the fry had nearly all left the pond, the 24th of May was fixed upon for the purpose of emptying it—to prepare for the hatching of 1855—when it was found that 335 smoults had remained behind the rest; they and 872 parrs were captured and turned into the river. These parrs, at the time, were thought to belong to the unsuccessful spawning of 1854, and were said to have been put into the pond by mistake. A few river trout, (salmo fario,) and some small eels were also found. The trout weighed from 1 lb. to 1½ lbs., and were supposed to have been mischievously put into the pond unknown to the keeper. We cut them open, and found them all gorged with smoults—so they must have reduced the numbers of the inmates daily.

The whole of the hatching of 1854 having now left for the sea, Mr Buist reports on it—"That the first of the fry that left the ponds as smoults in 1855 was on the 19th of May; the last on the 7th of June. No more left that year. The first of the same brood, which remained as parr all last season, assumed the smoult scales in April, 1856. The first division went off on the 28th of April, and the last on the 24th of May. In both years they went off daily in divisions, from the first to the last day. About 1300 were marked in 1855, and several returned as stated in my report. The number marked in 1856 was 300 with rings, and 800 with cuts in the tail. Taking one in each hundred as marked, it may be reckoned thus:—Left the pond in 1855, 130,000; do. in 1856, 100,000; total, 230,000. We shall anxiously look for the appearance of the silver rings." Although many grilse were reported to the superintendent as captured, having this year's mark, not one having the ring was amongst those taken. The grilses in 1856 were very numerous, and, had the fishermen taken more care to examine them, some perhaps might have been detected, but the greater part of them were unfriendly to the experiment. – However, the difficulty must be admitted of marking so small a fish as a smoult, (which seldom weighs much above an ounce when it is marked,) with any foreign substance that is likely to remain for any length of time attached to the fish, as its growth is so rapid, growing to the weight of 3 lbs. or 5 lbs. in six or seven weeks; besides, a silver ring is an attraction, no doubt, to the enemies of the fish while in the water, and therefore it is more likely to be devoured than any of its neighbours. A few of the fry that left the pond in May or June, 1855, were reported as having been caught this season as salmon—one of them being as heavy as 19 lbs.; another was taken by the late Col. Stewart of Dalguise while angling—the weight of which we did not learn. We may expect for some years to hear of large salmon being taken with mark of 1855.

The Stormontfield experiment having proved

so successful, many strangers from a distance were attracted to the spot. In January of this year the ponds were visited by M. De Ryan de Acuna, from Spain, and the Earl De Reus, a very wealthy Spanish nobleman of great influence in that country, in order to gain information prior to adopting the artificial method of rearing salmon to stock some of the rivers of that country with that fish. Although the projectors entertained doubts of many of the rivers of Spain being in a cold enough latitude for salmon, still they felt confident that those rivers that empty themselves into the Bay of Biscay would be found in every respect favourable for the purpose. We understood at the time that Mr Ramsbottom was engaged to commence the experiment, but we cannot say if success attended the proposal. Sir William Jardine, the well-known naturalist, M. Coste of France, and Professor Queckett of London, took much interest in the experiment at this time, and visited the ponds frequently. We have to acknowledge our thanks to Sir William for enabling us to distinguish the parr of the salmo salar, salmo trutta, salmo eriox, from the parr of the salmo fario, or common trout. Amongst a lot of thirty-one parrs, Sir William picked out the salmo fario, and showed that all the parrs of this species had the extremity of the second dorsal or dead fin fringed with orange, besides other marks about the head; but the fin mark is the best, and is easily distinguished. Being anxious to be able to detect the parr of the salmo trutta and salmo eriox from those of the salmo salar, we asked Sir William if he could give us any mark equally clear, but he frankly acknowledged that he was unable to detect any outward difference in the parrs of these kinds from that of salmo salar.

Although it may seem strange, it is not the less true, that, at this stage of the experiment, notwithstanding that every thing had been successful, the subject received more attention from persons at a distance than from those in the neighbourhood. The facts in the history of the salmon that had now been proved beyond a doubt, were again and again disputed by those who would not take the trouble to investigate the matter for themselves, but who still followed the errors of others, and were determined to retain them, because they did not square with their previously received opinions on the subject. Nothing daunted, however, Mr Buist, and the few friends that had taken an interest in the experiment from the first, kept on their way. Facts, not theories, in the history of the fish were what they desired. They were neither committed to the conclusions arrived at by Shaw nor Young, and ultimately were enabled to throw much light upon the previous experiments of both of these naturalists. Shaw, as we have stated, had insisted that the fry would not become smoults at the age of one year or thirteen months, while Young as confidently asserted that all of them would—the experiment proving both to be partly right, both partly wrong; one-half, as near as could be guessed, going off the first year, and the other half (having the parr marks) still remaining in the pond. Until the parr takes on the smoult scales, it shows no inclination to leave the fresh water. It cannot live in salt water. This fact was put to the test at the ponds, by placing some parrs into salt water—the water being brought fresh from the sea at Carnoustie—and immediately on being immersed in it, the fish appeared distressed, the fins standing stiff out, the parr marks becoming a brilliant ultramarine colour, and the belly and sides of a bright orange. The water was often renewed, but they all died, the last that died living nearly five hours. After being an hour in the salt water, they appeared very weak and unable to rise from the bottom of the vessel which contained them, the body of the fish swelling to a considerable extent. This change of colour in the fish could not be attributed to the colour of the vessel which held them, for, on being taken out, they still retained the same brilliant colours. We have also taken ova which had been recently manipulated upon, and dropped it into sea water, which destroyed it almost instantaneously. Only a few of them becoming opaque, in the greater portion of them the yolk became shrivelled up and contracted. We have put smoults which have had on the scales for some time into salt water directly from the fresh, and they seemed in their true element. We have also turned smoults out of a vessel containing fresh water into a salt water pond, wherein the sea flowed and ebbed, and we watched them as they left the brackish water at the side and sought the salt water as it was poured in unadulterated from the ocean, which they did a few minutes after. These facts prove that until the parr is covered with the new scales, it is unable to live in salt water, and also that salmon ova cannot hatch in the sea. Every one who has handled a smoult must have observed that the smoult scales come off and stick to the hand, and that the parr scales and marks were still there, having been only covered by the smoult scales. The fish, when undergoing this process, changes its nature altogether, for instead of being a firm, wiry fish, as the parr is, it becomes soft and easily injured—in short, it seems to have undergone a new birth. When the smoult returns as a grilse, its scales come off with the slightest handling, and it is only when it returns as a salmon, or has been long enough in the sea, that the scales become rigid and firm.

EXODUS, 1857.

The fry of the hatching of 1856 having shown symptoms of becoming smoults this year, 1857, by the first week in April the keeper was on the outlook for the departure of the first shoal, which took place on the 12th, being a month earlier than the first exodus of the hatching of 1854; which we have no doubt was caused by the winter of 1856-57 being much milder than 1854-55. As the marking of the smoults of the previous hatching with rings had failed, Mr Paton, of Messrs Paton & Walsh, gunmakers,

Perth, contrived an instrument which punched out a small part of the gill cover in the shape of the Greek letter Delta, Δ; the mark was easily put upon the fish, which appeared not to be in the least injured by it. Mr Harvey again sent 300 silver rings of an improved construction—invented by George Anderson, Esq., Glasgow—which, to all appearance, gave hopes of their being found in the grilse on their return. The young smoults on their leaving were marked as follows:—

    About 270 with silver rings inserted into the fleshy part of the tail.
    About 1700 with a small hole perforated in the gill cover.
    About 600 with the dead fin cut off in addition to the above perforation.

Mr Buist published circulars offering a reward of £1 over and above the market price for the first grilse brought in; Ten Shillings each for the next ten; and Five Shillings for the second ten grilses.

On the 3d of June a grilse was caught in Lunan Bay, near Montrose, and sent to Mr Buist for inspection. We examined this fish, and although the part that was punched out of the gill cover was filled up, still the covering was so thin that when the gill cover was held up to the light and looked through, the mark was quite visible. The dead fin had been cut off, and it was this mark that drew the attention of the fishermen, for the punched out mark of the gill cover could only be detected after careful inspection, which rendered this mark almost useless. This was to be regretted, as it was the mark easiest put on the fish and that injured it least. Although the mended covering of the punched out hole was of a lighter colour, still the difference was not sufficient to attract the eye of the fishermen, and from this reason it was a failure, as few could be expected to be reported having this mark. This fish weighed 3½ lbs., and was sent to Sir William Jardine, who wished the first fish sent to him having any of the marks. It was generally thought at the time that the gill mark would not close up, but future experimenting proved that this was not the case. Lunan Bay is said to be the place where the river Tay entered the sea before the historical period—the chain of lochs through Strathmore almost placing this beyond a doubt; for were the rocks which cross the river at the Linn of Campsie joined together, as they must have been at one period, the Tay would still enter the sea at this bay. It is also singular that in this bay a number of the salmon marked by the Duke of Athole have been caught; and the fishermen acknowledge that the most of the fish caught at this place are Tay salmon.

By the end of the first week in June, as in former years, about one-half of the fry in the pond had left as smoults at the age of 13 or 14 months, and the other half still remained as parrs, although, as on former occasions, there was no obstruction to keep them from entering the river, and going off with the smoults. This repetition of the same phenomenon convinced every candid observer that this process took place in the river as well as in artificial rearing; but we shall see afterwards that this anomaly will account for much that was before considered dark in the history of this fish.

On the 13th of July, of this year, 1857, there were taken at a fishing station near Montrose 155 grilses, and one of these was a Stormontfield fish. The tacksman wrote to Mr Buist, "that he had sent away a marked grilse about 4 lbs. weight. It had a small round hole through the gill-cover, and a piece of the dead fin cut off. Does it answer any of your Tay marks?" This tacksman evidently did not know what was this year's mark; and, as he has described it correctly, there can be little doubt of it being one of the marked smoults of this year. In this instance, the hole had only partially closed up. In addition, a number of grilse bearing the mark were caught and sent to the superintendent, but in every instance the mark in the gill-cover was filled up. The want of the adipose fin called the attention of the fishermen to the fish, and, when closely examined, the excision, although covered over, was plainly discernible. On the 28th of July a fine grilse, weighing 7 lbs., was caught at Tappie station—one of Lord Gray's fishings—which was at once recognised by the fishermen as a pond-marked fish. Mr Buist sent it to Sir William Jardine, who stated that he was quite satisfied that it was a pond fish of this year's marking, but he regretted the filling-up of the mark made in the gill, as the fish would have been more easily recognised had the mark not closed up. At Sir William's suggestion a number of smoults were marked by the same instrument as those that were turned into the river and put into the filtering-pond, when it was found, on capturing them some months afterwards, that the hole had closed up—clearly proving that Nature makes every effort to perfect this covering to protect the gills. No ringed fish was caught, which was much regretted; and we question if ever a foreign substance can be inserted into a smoult—at least, while in the fresh water—that will remain for any length of time attached to the fish after it goes to the sea.

On the 1st of October, 1857, the celebrated M. Coste, Professor in the College of France, and whose successful experiments in the artificial breeding of fish and re-stocking the rivers of that country, are well-known, paid a visit to Stormontfield. The Professor spoke in high terms of the manner in which the experiment had been conducted, and carefully inspected the fry in the pond. M. Coste recommended that some more small ponds should be added to the one in operation, so as to enable the experiments in the natural history of the salmon to be thoroughly investigated, but this suggestion, as it involved expense, was never acted upon; for although the rental of the salmon fishing on the Tay is, at the present time, about £14,000 a-year, some at least of the proprietors object to the small sum—about £50—that is required annually to keep up the present establishment.

FOURTH SPAWNING, 1857.

On the 12th of November this year, 1857, the artificial spawning of the salmon commenced, and finished on the 2d of December, during which time 15 salmon and 14 grilse were spawned, giving about 150,000 ova. Besides, in his report to the proprietors, Mr Buist states "that 89 salmon and 78 grilse were landed in the net, but not being fully ripe, were, with the greatest care, returned to the river, after being marked with copper wire. None of the fish seemed to suffer in the least, and all went away in a very lively state."* Peter Marshall manipulated, and 188 boxes were filled. Mr Shaw of Drumlanrigg had already stated that he had successfully experimented with the milt of the male parr shed on the ova of the female salmon, and this season it was resolved that this experiment should be repeated. A female salmon, of about 20

---

* The only one reported as retaken with this mark was a salmon, 13½ lbs., caught in Largo Bay, on the 6th of September, 1858, the wire of which was sent to me.

lbs. weight, was caught, in excellent spawning condition, and the keeper having provided, *from the pond*, a number of male parrs, artificially reared, in a similar state, the ova was fecundated by their milt. Two parrs were required for every 400 or 500 ova. The milt of the parr appeared to have the same effect on the ova as that from the male salmon, and it was expected, therefore, that they would hatch. They were put in boxes by themselves, and carefully attended. M. Coste, when visiting the ponds, recommended that experiments should also be made upon the ova of sea-trout (salmo trutta), impregnated by the milt of salmon (salmo salar); but it was thought, as there were no additional ponds to receive the fry, it would be better not to give occasion to any one to say that the fry had been mixed; and this principle has been followed out all along in the Stormontfield experiment, as nothing but the salmo salar has been experimented upon. The keeper, however, furnished the writer with a few ova taken from a female sea-trout, impregnated with the milt of a grilse, and, in 60 days, by the syphon process, the water at a temperature of 44°, the ova were hatched. They appeared to be weakly, larger in the head and longer in proportion than the fry of the salar in the same stage. We had no proper way of keeping them at the time, but they lived for 40 days. It would be interesting, if not advantageous, to the proprietors of the salmon fishings to carry out this experiment, so as to ascertain if such a fish would live and propagate its kind. As M. Coste recommended, experiments should also be made with the ova of the salmo trutta, or sea-trout, and also with the ova of the salmo eriox,—if it were only for the purpose of ascertaining by comparison if there is any recognisable difference in the parr and smoult of this fish and that of the salmo salar or salmon. But we are afraid that the expense of the additional apparatus will deter the Tay proprietors from prosecuting these researches, as hitherto they have shown little inclination to proceed further than they have already done.

HATCHING OF 1858.

The ova of this hatching, having had a very favourable season, burst their covering on the 2d of March—there were hardly any addled eggs. On the 17th of April the first smoults belonging to the previous hatching made their appearance in the marking box on their way to the sea. The mark this year was cutting off the dead fin and a portion of the upper part of the tail. Mr Harvey of Glasgow again furnished 100 silver rings, which to all appearance seemed to promise fair, but none of these so marked smoults were ever reported as having been taken. About 50 of the fry were also marked by copper-gilt wire inserted into the fleshy part of the tail, and likewise in the back behind the dead fin with a like result. When the pond was emptying it was discovered that there were from 50 to 60 of the fry that were still parrs, about two dozen of which were caught and put by Mr Walsh into the filtering pond; and we put a dozen into a small run of water on our premises, and these fish did not become smoults till next April. This is a curious fact, for in size these fish were equal with those that left the pond as smoults. They increased only about an inch in size by next spring, 1859, but they had become smoults, and by the spring of 1860, they had not grown more than another inch. It was interesting to watch the change of colour which came over them during the spawning months, but we believe none of them were ever dissected so as to ascertain if the females contained ova. During a flood in the lade which supplied the filtering pond with water, they managed to escape into the lade, which was much to be regretted. We shall have occasion to speak afterwards of one of those in our possession.

The spring of 1858 was one of the most backward experienced for many years, and the fry in the pond were late of becoming smoults. On the 8th of May we marked 25 with the silver ring behind the dead fin, and about 50 with gilt copper wire; but the mark this season was cutting off the dead fin entirely, which fin has been proved does not grow again, hence no other mark could be depended upon. The smoults this year were larger and finer grown than on any previous occasion—specimens may be seen of them in the superintendent's office, and also in the museum of the Literary and Antiquarian Society of Perth. The silver rings are no doubt an attraction to the enemies of the fry; for in this instance, when the marking sluice was opened, and the fish allowed to escape into the river, a pike or large trout gave a plunge, and the keeper exclaimed, "There goes one of our smoults." We dulled the rings after this, but we have no doubt that a small fish carrying a silver ring in its back must be very attractive to pikes and other enemies.

The hatching of 1858 turned out very prolific, for the lower canal contained as many fish, as, if allowed to come to maturity, would have covered the whole rental of the Tay for one year; and when it is considered that these fish would meet with no accident in the pond—as ten per cent. of deaths has not happened in any year since the ponds were established—it will be easy to see the value of artificial rearing. At this time a visit to the ponds must have been very interesting to the naturalist, for the fry of salmon, the fry of grilse and salmon, the fry of grilse, and the fry of salmon and parr —kept in separate boxes—might have been seen swimming about, and compared. On the closest inspection, no difference was perceptible either in the form, colour, size, or markings of any of these fish. There were larger and smaller fry to be seen amongst all these hatchings.

Very few of the smoults of this exodus—May, 1858—were reported as grilse to Mr Buist as taken in the net, the fishermen not giving themselves the trouble; but a number were captured by anglers with the undoubted mark, namely, the dead fin cut off, and a piece off the upper part of the tail. The late Professor Queckett of London, while angling in the Tay, near the ponds, in the month of September, caught a fine grilse of 6 lbs. weight, as he reported, with the undoubted mark of 1858; and many more taken with the rod were reported to us. None of those that were marked by the silver rings or gilt copper wire were ever known to have been taken. The boxes were not re-stocked in 1858, as they can only be re-stocked every alternate year, from there being only one pond, as previously stated, and as it takes two years to bring all the parrs into the smoult state, and even then a few are still parrs.

EXODUS OF 1859.—HATCHING OF 1858.

On the 14th of April of this year, it was evident that many of the hatching of 1858 had become smoults; but it was not until the second week in May that the first smoult left the pond of its own accord. Mr Buist was convinced that it was useless to mark the fry in any other way than by cutting off the dead or second dorsal fin, as all other marks had hitherto failed. The number marked this year was 506.

The first grilse reported as taken with this

year's mark was caught at the North Inch station. Mr Speedy, tacksman, who has all along shown much interest in the experiment, and is therefore an exception to most of the other tacksmen, apprised Mr Buist of the circumstance, who was so kind as send us notice of its capture. In company with Mr Buist we examined the fish, and found that it weighed $3\frac{1}{4}$ lbs. Mr Speedy cut off the dead fin from some grilse that were lying beside the pond fish, to show that it was impossible to deceive, by any recent abstraction of the fin, for the pond-fish had the part covered by skin and scales. The fish, when marked as a smoult, weighed not more than an ounce, or an ounce and a-half, at the most, had increased in weight 3 lb. 3 oz. in six weeks. The take of grilse this season was unprecedented, for on Monday, the first of August, one tacksman alone caught upwards of a ton weight of salmon and grilse. Counting them was out of the question, far less looking for pond fish amongst them. They were shovelled into boxes as fast as boxes could be made, weighed, and sent off to London; but we may infer that had they been examined, many pond fish would have been found amongst them.

SPAWNING OF 1859.

This year, 1859, the first net was shot at the mouth of the Almond, in order to obtain spawning fish to re-stock the boxes. The following is Mr Buist's report of the fish taken; 150 boxes were filled:—

	Taken Salmon.	Grilse.	Male.	Female.	Ova.
Nov. 15,	10	4	0	0	
" 17,	7	9	3	6	60,000
" 21,	10	0	0	0	
" 23,	20	3	2	3	30,000
" 25,	16	2	1	1	10,000
" 26,	38	5	3	5	60,000
" 28,	1	1	1	1	1,000
" 30,	46	2	3	8	40,000
Dec. 2,	44	4	3	5	30,000
" 3,	7	0	1	1	4,000
" 13,	16	1	2	2	20,000
	297	31	19	32	255,000

Nearly all the ova that were deposited this season were salmon ova fecundated by the milt of male salmon,* in order to test, on a larger scale than was tried in any former experiment, if the proportion of fry remaining parr for two years would, under these circumstances, be still the same; and the result has shown that there is no

---

* There was only one female grilse spawned on the 28th November.

difference, either in quantity or appearance of the fry, after the first year, from fry raised from grilse alone, or salmon and grilse conjoined, or even salmon and parr. All take the same time to hatch, and in equal proportions arrive at the smoult state at the same time.

HATCHING OF 1860.

The winter of 1859 and 1860 was very unfavourable for the hatching of the ova, and fears were entertained that a good portion would be destroyed by the severity of the weather; but although this retarded the young brood, they turned out as numerous as on former occasions. The keeper reported the first ovum to have been hatched on the 10th of April. The first of the ova having been deposited on the 17th of November, it had taken 145 days to hatch, which is about a month longer than it would have required had the season been mild and good. The keeper says that 128 days was the longest previous time of hatching since the experiment commenced. He also stated that the temperature of the water, after the deposition of the ova, never was higher than 36 degrees. The two-year old fry in the pond at this time were still in the parr state, and it was not until the beginning of May that they commenced to put on the smoult dress. By the 26th of this month the pond was emptied to admit the young fry that were in the boxes and canals. As on former occasions, when emptying the pond, it was discovered that from 50 to 100 of the fry were still in the parr state. A few of these were put into the filtering pond, and did not become smoults until 1861, the rest were put into the river. 670 smoults were marked this year. The mark was cutting off the dead fin and a part of the lower part of the tail. In order to prove whether the fins grew after having been cut, Mr Buist again desired the keeper to cut off the dead, or second dorsal fin, and part of the tail fin, and also to mark the gill cover with Mr Paton's pinchers, on a number of the two-year old parrs that had not assumed the smoult scales, before he put them into the filtering pond, and the result was that all the marks became obliterated except the marking by cutting off the dead fin, which never was replaced. On the first of March, 1860, Robert Nicholson, Esq., who rented the rod-fishing at Stanley, killed a fine salmon with the mark of 1855. It weighed 22 lbs. A good many grilse were reported as having been caught this summer with the mark, and also a number of salmon of the previous hatchings; but we believe no regular account was kept—the fishermen still being so much against the experiment.

EXODUS OF THE FRY, 1861.

The winter of 1860-61 having proved so severe, it was the third week in May ere the keeper reported that the smoults in any great numbers were descending to the river. The fry of this year being from salmon ova, much curiosity was felt to discover if the smoults were larger, or had any difference in appearance from former hatchings. On the most careful inspection no difference could be detected, for although many were fine large smoults, not a few had not grown larger than minnows, and had the parr marks well defined. The question, then, Whether all the fry that became smoults the first year were the produce of salmon alone? is now set at rest, and we are at as great a loss as ever to account for the difference of age at which parr of the same hatching become smoults. We suspect it is a law in the history of the salmo salar that will never be unriddled. The marking this year was omitted.

The same remarks with regard to the takes of marked fish of former seasons may be applied to this. No marked grilse could be looked for, but we heard of several salmon of former markings having been captured, both with net and rod, though not officially reported to the superintendent. Few, if any, of the fishermen have any doubt about the matter when they capture a pond fish, but they will not take the trouble to report; and until that is done in every case, no correct data can be formed of what the experiment has done and is accomplishing in increasing the number of salmon in the Tay. The matter, we think, rests with the proprietors of the fishings themselves, who, on letting their fishings, could bind down the tacksmen, so that their men would be obliged to report on all pond fish taken. Without something be done towards arriving at more than a mere guess, it is next to useless going on marking the smoults.

We have now described the experiment as it proceeded, up to the present time. The hatching boxes have again been re-stocked for this year, 1862; but, from the flooded state of the Tay during the spawning months of 1861, the quantity of ova obtained was small. The keeper, however, has collected a considerable amount from the beds which were left almost dry when the river fell low, by this means securing the ova which otherwise would have been lost.

The question that now remains to be considered is,—Has the artificial propagation, even on the small scale that has been carried on at Stormontfield, been of advantage to the fishing proprietors on the Tay? We have no doubt about the matter, for on referring to a statement of the rental of the Tay published by the proprietors themselves, we find that in the year 1828—the year of the passing of Home Drummond's Act— the rental was £14,574. It gradually fell off every year afterwards until 1852, when it reached the minimum, amounting to £7,973 5s. In 1853 the artificial rearing commenced, and in 1858, when the statement was printed, the rental was £11,487 2s 5d; it has now, 1862, reached what it was in 1828. We are aware that other reasons are given for the rise in the rental, such as the extra price of the fish in the London market; but we should like to know how it happens that all the other rivers in Scotland (with the exception perhaps of the Sutherland rivers) which have the same market for their fish, have, since 1852, had a lower rental instead of an increased one. It is true that in 1853, 1854, and 1855, the proprietors of the Tay voluntarily agreed to close their net fishings upon the 26th of August instead of the 14th of September, and that, in 1858, an Act was passed legalising that agreement, which has no doubt done much to increase the number of fish in the river; still we are of opinion that the great rise of the rental in nine years cannot be accounted for in any other way than from the pond-bred fish, and if the fishing proprietors would see to their own interests, they would have many acres of breeding boxes and ponds made to rear and preserve their young fish. At present, however, their rental is in the ascendant, and they are contented; but, should a reverse take place, we would then see artificial propagation much in favour.

In corroboration of these remarks, and of the advantages to be derived from the artificial culture of salmon, we have been favoured with the following remarks upon the subject by Mr Thomas Ashworth, the originator of the experiment, and we are at liberty to say that his conclusions have received the sanction of Mr Ffennell, one of the commissioners on the Irish fisheries. Mr Ashworth and his brother are the proprietors of the Galway fisheries, which, with the districts of Lochs Mask and Carra, comprise an area of 30 miles long by 10 wide, and which receive the waters of some of the finest tributaries for the purposes of propagation. Mr Ashworth's experiments in artificial propagation in that quarter of Ireland are too well known to need repetition; and we are glad to learn that he is still carrying on his experiments, as, this season, he has deposited at the least 659,000 salmon ova in the tributary streams of Lochs Mask and Carra, where salmon ova were never seen before, although both these lochs are connected with Loch Corrib, which abounds with salmon—a natural barrier of rocks preventing the ascent of the fish. This barrier Mr Ashworth is about to remove, as not a single fish has been observed to pass up; we have no doubt, however, but that the smoults reared in the upper tributaries will seek their way back to their native stream, and in a few years stock these lochs and rivers with this fine fish. We shall be glad to hear of the success of this experiment, which will tend further to elucidate the history of the salmon. Whether this gigantic experiment succeeds or proves a failure, Mr Ashworth deserves the thanks of the country for his labours. Mr Ashworth, in advocating the adoption by salmon proprietors of artificial culture, states it as his opinion, and that of Mr Ffennell also, that not above 1 in 6000 salmon ova, deposited naturally in the bed of a river, arrives at the grilse or salmon state and becomes marketable. Their reasons for coming to this conclusion are, "that, as 12,000 fish were caught in his river in one year on their way to the spawning beds, these fish must have ascended the river the previous year, and also the year previous to that, and that at least half that quantity must have gone up for the third or fourth years, consequently that more than 24,000 fish, of various ages, ascended the river annually, and taking one-half to be females, we get 12,000 fish of various sizes, from 6 to 12 lbs. each, and producing, as is well known, at the least, 6000 ova each fish, would yield 72,000,000 of ova deposited, or 1 from 6000 ova deposited naturally in the river. Stormontfield ponds, we know, have hatched 300,000 ova in one year; this quantity hatched in a natural way, in the bed of the river, would have produced one matured fish to each 6000 ova so deposited. Then arises the question,—how many smoults were raised and liberated out of the 300,000 eggs? We think we may calculate upon 30,000, or 1 in 100 of the quantity sent to the sea. We have no means of ascertaining the quantity that did return; but one thing is certain, that we are enabled, by artificial means, to send to sea a vast deal more smoults than would have gone had the same quantity of ova been hatched in the river, where it is exposed to all its enemies for the space of fifteen months. We also know that, just in proportion to the quantity of smoults migrating to the sea, an increased or decreased quantity return to the rivers. We know, also, that the smoults go to the sea in millions, and the only return we get as grilse and salmon is the comparatively few thousands that we catch. Experience proves to us that the produce of 12,000 salmon would be at least 72 millions of ova, and yet, from this vast quantity of seed, we have returned as salmon only as many fish as are produced from the ova of two or three fish, or, say one fish for 6000 ova. It is evident that the loss in the rivers is great; and of the amount that takes place in the sea we have no means of ascertaining it, but we know that it is beyond calculation, the enemies of the salmon are so numerous."

NATURAL HISTORY OF THE SALMON AS LEARNED FROM THE STORMONTFIELD EXPERIMENT.

We shall now proceed to state what the Stormontfield experiment has taught us of the Natural History of the Salmon.

The parr, that little active fish that is to be found in all salmon and trout rivers, was generally considered (not forty years ago) to be a fish *sui generis;* and even at the present time, by many persons who have not given themselves the trouble to inquire into the matter, parrs are believed to be full-grown, perfect fish. Learned judges have been puzzled on this question; and how could it be otherwise, from the conflicting evidence given (witness the Dunblane case) before them by fishermen who had been all their lives engaged in salmon fishing, as to what a parr is? Hogg, Shaw, and Young, were the first that proved that, at any rate, the fry of the salmon, salmo salar, was a parr before it became a salmon;

but the Stormontfield experiment, having been done on a more extensive scale, and repeated from year to year, with the same results, has settled the question. Almost all fishermen and superficial observers believed, because there were parrs to be found in salmon rivers at all times, the parr was a distinct fish; but the experiment has shown that this error was likely to arise from one-half of the hatching of a fish's ova becoming smoults the first year, the other half remaining in the river for another year as parrs, and in some few instances for three years, ere they are ready to migrate. These observers, also, never took it into account that the fry of the salmo trutta (sea trout), salmo eriox (bull trout), salmo fario (common trout), are all parr in their young state—all marked with transverse bars in nearly the same manner, the difference between them and the parr of the salmo salar, or salmon, being so slight that, except to the naturalist, little or no distinction is perceptible—hence the number of fry that was constantly destroyed in the parr state must have been great—it was only when migratory parrs became smoults that they were protected by law. We also learn that the ova of salmon, at least, are not fecundated until they leave the fish, and that the male parr is as fit to continue its species as the adult male salmon, but no female parr has yet been discovered with roe developed. The experiment has also established the fact that there is no difference in the length of time taken to hatch, or the appearance of the fry after hatching, up to the smoult state, between the fry of salmon, grilse, salmon and grilse, salmon and parr, or grilse and parr.

It has also been ascertained that the fry of the first year that assume the migratory dress, are composed of both sexes in nearly equal proportions, and it makes no difference whether the fry be reared from salmon or grilse, or salmon and grilse, or salmon and parr, or grilse and parr. Why those that remain behind for another year do so, and a few no doubt of each hatching for a year more, we cannot tell, but such is the fact; and the best reason we can venture to give is, that by this means the river has always fish in it, that will migrate at least a month sooner in the spring than the fry of the first year, and also that male parrs will always be at hand in the river during the spawning months in a fit condition to supply the want of male salmon, when that occurs, which is a wise provision in nature, as many females in small and distant tributaries might be without a mate, if there were no parrs, male parrs having been proved to be in a breeding state at that time.

The question of salmon spawning in the sea has also been settled—no salmon will spawn in the sea, if it can help it—as salt water destroys the ova. The experiment has demonstrated the practicability of rearing salmon artificially, fit for the market, within twenty months from the deposition of the ova—and the great value of artificial production—as it is ascertained that not above ten per cent. of the ova deposited in the boxes is lost, and not above twenty per cent. additional, but arrives at the smoult state and is sent into the river—(the keeper's report is much under this)—whereas it is generally supposed that not above 1 in 1000[*] of those naturally deposited in the river ever arrive at the smoult state—being the prey of fowl, fish, insects, and many other enemies. It has been also noticed that not a few of the ova deposited in the natural way miss the fecundating milt, and are lost; and when we take into

---

[*] Mr Ashworth and Mr Ffennell, as previously shown, estimate 1 in 6000.

account the great quantity that is deposited during floods on places that are left dry when the river falls in, and also the numbers of redds that are sanded up by large spates, we need not wonder if only 1 in 1000 should ever become marketable. The keeper this season—January, 1862—has been employed for many days carrying the spawn which has been deposited in places of the river which have been left nearly dry, by the river falling in, and spawn sanded up by the spates, and placing it in the boxes to be artificially reared. Hence the obvious advantage of artificial breeding to rivers that have been overfished, or to those that have been destroyed by poaching and obstructions on the river to the ascent of the spawning fish, etc., etc.

That the smoults return again to the river in which they are reared, has also been proved by the number of the marked grilse which have been caught in the Tay since the experiment commenced. If all the smoults which left the pond since the first exodus in 1855 had been marked by cutting off the second dorsal fin, we are quite satisfied that scarcely a day during the fishing season would pass without grilse or salmon bearing the mark being captured;

but the marking of two or three hundred thousand fish is a difficult process, seeing they leave the pond in such immense shoals; however, undoubtedly many more might have been marked, if more hands had been employed to do it. One great difficulty in marking is, the shoals generally leave during the night: it is true the sluice, which is always open, might have been kept shut during that time, and only opened during the day; but we would not approve of keeping back the fish at any time when willing to go. We understand Mr Buist is to arrange for as many as possible being marked this spring.

The experiment, also, has proved that the marked grilse of one year return as salmon the next; and we think it has also proved that all the smoults of one year do not return the same year as grilse, the one-half returning next spring and summer as small salmon. This idea was first suggested by a correspondent in the *Perthshire Courier,* who says, "That the habit of the salmon species, in adopting a double or divided migration to the sea as smoults and from the sea in their later stages, has only been known within these few years. It was discovered, in regard to smoults, by means of the Stormontfield ponds, and it is a far more important discovery than the parr question, which was settled by these ponds. People have been endeavouring, by all sorts of theories, to account for this divided migration of the smoults; some say that the males went and the females stayed, others that the produce of old salmon went and those of the grilse remained; but the fact is, simply, that it is a habit of the fish. A little reflection enables us to see that it is not confined to the first migration, but is repeated on the return of the smoults as grilse one year and as salmon the next, and on the return of the old fish after spawning—one-half coming back the following summer, and the rest not returning till the spring of the succeeding year. This habit of the fish has enveloped the subject in difficulty, but a clear apprehension of its truth will unravel the mystery." The writer also says, "We may only further remark that Mr Mackenzie of Dundonnell, in his Appendix to his father's book, lately published, in attempting to clear up this matter, has missed the blot, and has started a grievous heresy, viz., that salmon and grilse are different kinds of fish. Of course, this is out of the question altogether, for the same lot of ova at the Stormontfield ponds has produced smoults that were caught marked as grilse the first season, and salmon the second; but his idea may be taken as a corroboration of what we have stated, to this extent, that all the smoults do not return the first summer, but that those that are to return as salmon remain in the sea from fifteen to eighteen months; but that this is a provision of nature, or natural habit of the same species of fish, we affirm to be the true solution of the riddle, without resorting to the desperate expedient of maintaining, as Mr Mackenzie has done, that salmon and grilse are different species!" We quite agree with this writer, as his idea is the only one that will explain why clean salmon, having only a thread of milt or roe in them, are to be found in salmon rivers in the months of December, January, and February. These fish are falsely called barren fish, but this is not the case, as many have been dissected by Mr Ffennell, and in his examination before the Lords' Commission (see Appendix) he proves that these fish were not barren, but had ova perfectly developed, although as small as a mustard seed. He says, "These fish do not spawn until November and December following, remaining ten or twelve months in the fresh water, their ova developing until ready to spawn, and although they get very much coloured by being so long in the fresh water, and not a very good-looking fish to send to the market, it is a right good fish to eat." We were so fortunate as to capture one of these fish with the fly, in one of the Sutherlandshire rivers, and, as we had never seen a salmon so much coloured in the summer season, we imagined it was a fish near the spawning. We were about to return it to the river, when the keeper, who was with us, said it was a fine clean fish, and put it in the creel. It turned out so, and, as Mr Ffennell says, "right good to eat." Mr Ffennell also remarks, "I think it as clear as possible, that the object of the law that governs them is to cause the fish to distribute themselves throughout the whole length and breadth of the river." These fish, by ascending the river so early to spawn, and having the spawn so slightly developed, are strong, and able to overcome falls, and penetrate to the extreme feeders of the river, whereas those fish that do not leave the sea until heavy with spawn, could never overcome falls, or ascend any considerable distance. Now, we think that the anomaly ascertained to occur in the migration of the fry explains all this, and accounts for the seeming irregularity in the history of the fish, which is still carried on in all its stages of growth, as we have previously shown when in the grilse state, and this knowledge is again the result of the Stormontfield experiment. Further, the experiment has proved that all the smoults do not return the same season as grilse, but that not a few of them remain in the sea, and do not return to the river until the spring and summer of next season, not as grilse, but as salmon of from 4 to 10 lbs. weight. The smoults with the mark of the first exodus which were caught were not all captured as grilse, but some of them were taken the following spring as small salmon, which explained what had never been accounted for—namely, the cause of small spring fish. Every fisherman was familiar with the appearance of the spring salmon—they generally succeeded the large strong early runners; but a few are always caught running up along with them. We saw one taken on the 1st of February this year, 1862, at Burnmouth station, which weighed only $5\frac{1}{2}$ lbs., which was, although a salmon, only a late grilse of last year. The scales of this fish when taken out of the water were not easily removed by the hand, like the scales of the grilse, and the tail was not so forked.

ON THE ADVANTAGES OF THE ARTIFICIAL PROPAGATION AND REARING OF THE FRY OF SALMON UNTIL THEY REACH THE MIGRATORY PERIOD.

Some naturalists, when they have discovered a new species of fish, are quite satisfied after they have arranged and classified it according to some well-known system, or perhaps a new one invented by themselves. They give the creature their own name, with a Latin or Greek termination, and hand it over to the public. But how much is learned of the history of a fish by an anatomical knowledge of its structure may be seen by the little that is yet known of the true history of the herring (*clupea harengus*), or what is yet to be learned of the history of the salmon. Prying into details, and searching out the habits of some well-known individual of the piscatory family, may appear dry work to the ardent naturalist; but the labour in the end would be more useful in an economical point of view. From the discoveries which have been made with regard to what are the real enemies of the young salmon in its natural haunts, we are prepared and have the means in our power, to save it from them until it reaches the smoult state, ready for migrating to the sea.

That much destruction of smoults takes place in the salt water, there can be little doubt—its enemies there are as numerous and more voracious than even those of the river; but when the animal is so destroyed in both its habitats, it is marvellous that so many are left, and were it not for the enormous produce of this fish, it would soon become extinct. Man is another great enemy—not so much, perhaps, from the numbers that are caught by him and sent to the market, but from his barring up the rivers to the ascent of the fish to its secure spawning grounds, by fixed engines of all kinds, weirs, dam dykes, etc., etc.; for unless the salmon can have a free run in spring and summer—when the fish is at the strongest—it is unable, when full of spawn, to surmount the natural obstacles which it meets with in the river, and hence, instead of the small mountain streams becoming the nurseries of the young fry until the migration period of its existence arrives, the gravid fish is compelled to deposit its ova on some ford of the main stream, where they are frequently sanded up to the depth of many feet by the winter floods, and consequently lost. To obviate this, and to allow a proper stock of spawners to reach the higher feeders of a large river, the weekly closetime should at the least be forty-eight hours. The natural enemies of the salmon in the sea, man has no power to curtail. This is not the case, however, in the fresh water, for he can increase or diminish by his protecting or destroying them, with the exception of the insect tribe, over which he can have little or no control. To diminish the number of insects in a river, even if it were possible, would be unwise, as their larvæ and themselves are the principal food of the young fish; but far more of the ova of the genus salmo are destroyed by the larvæ of insects than by all other enemies put together. The pike, the river trout (salmo fario), and other foes to the increase of salmon, if salmon proprietors were alive to their own interest would soon be diminished in our salmon rivers; instead of that, the enemies of the salmon have increased for some years past in most of our salmon streams. This fact cannot be doubted, for at the present time our salmon rivers are shut to all but the favoured few who have obtained permission from the proprietor, or who can afford to pay dearly for the privilege. The angler who contented himself with returning home in the evening from a salmon river with a basket full of trout and pike, is no longer to be seen; in short, the salmon is now the only fish that is angled for in a river of the kind, and, as a natural consequence, his destroyers have multiplied amazingly. But how are the insects to be overcome? We know no way but by artificial propagation of the fish. The mayfly (ephemera), and a host of other flies, drop their eggs during summer on the surface of the water from which they have just arisen, which eggs, from their specific gravity being heavier than water, sink to the bottom, where they lie until they are hatched, after which they become, when encased in their shell, the most deadly foe to the increase of the salmon, by devouring the ova in millions. They are assisted in this process by water-beetles, water-fowl, and all the other tenants or occasional frequenters of the stream; besides, great loss is sustained by the sanding up of the ova after deposition on the fords by winter floods, as has already been alluded to.

Previous to the introduction of fixed nets on our coasts, and when the net and coble was not fished so closely and unceasingly as it is at present, multitudes of strong early salmon were permitted to ascend the rivers and fill our Highland streams with spawners, which streams contained a fair share of marketable fish during the summer months. The case is now altered, and those who know these feeders will bear us out when we say that there is at present not one fish for an hundred that used to be found in these upper waters. If this is the case in the best spawning grounds of the salmon, how can we expect any other result than a falling off in the number of fish taken in the river? All the fish which are taken in stake or bag nets will never make up the deficiency; and if the law is to remain as it is at present, and if these nets become legalised, and if the weekly close-time is not extended, the sooner our lower salmon proprietors begin artificial propagation on an extensive scale the better. Upper or Highland proprietors of fishings can have little or no interest in artificial propagation, for very few fish at present are allowed to visit their waters, and they cannot be expected to breed fish artificially for the lower proprietors. As it is, they have small inducement to protect the few spawning fish that are permitted to reach their streams.

There cannot be any doubt but that artificial propagation will greatly increase the number of salmon in any river, when it has been proved that not above thirty per cent. of the ova deposited artificially is lost, and about 1 in every 100 return as grilse: whereas, Messrs Ashworth and Ffennell have calculated that not 1 in 6000 deposited in the river ever becomes marketable. This immense difference will not be wondered at when we consider how much the ova deposited in the beds of our rivers are exposed to dangers of all kinds, and to enemies innumerable, whereas those deposited artificially are carefully watched and protected from enemies of every kind, and reared with food upon which they thrive, and rival the fry of the same age in the river in size and weight.

Nor is artificial rearing an expensive affair. The Stormontfield ponds cost, at first, about £600, and they require about £50 annually for their support. They send, every spring, to the sea from 150,000 to 200,000 smoults, and if 1 in 100 of these return to the river as grilse, and are caught, as Mr Buist's statement has proved, the experiment must be a profitable one, when we consider the small outlay. Thirty or forty spawning fish are required to fully stock the Stormontfield boxes. These fish are taken from a ford at Almond Mouth, where we have witnessed many dozens of pairs spawning at the same time; and we are convinced that where too many fish are compelled to spawn on the same ford, for want of water in the river to enable them to ascend, a very large per centage of the spawn is lost. We have again and again seen a pair of fish busy at work, the female dropping her ova, and the male shedding his milt—the female alone covering them with the gravel, by digging into the upper part of the hole she had made, allowing the stream to carry down the gravel. This operation takes some time to perform; but we have more than once observed that, some hours afterwards, another pair of fish were as busy digging away in the same spot at a similar operation. Now there can be no doubt but that the ova deposited by the first fish was nearly all destroyed by the second pair working on the same ground—for we saw numbers of trouts and whitlings (the grilse of the salmo trutta, or sea trout) busily engaged devouring the spawn as it was turned up. When fish are kept from running up the river, either by late fishing, or want of water, they get gravid, and must deposit their spawn on the first ford that presents itself; hence we have witnessed at Perth Bridge—where the tide rises at spring tides eight or ten feet—as many as fifty fish spawning at the same time. When this occurs—which it does every dry season when the fish are unable to ascend—artificial rearing would save millions of fish. Or, when the season has been wet, and the fish have been tempted by the height of the flood to deposit their spawn in places of the river which are left dry when it falls in, millions of ova can be gathered up and reared artificially which would otherwise be lost. The French Government, under the superintendence of M. Coste, has taken up the subject of pisciculture in earnest, and, as is well known, great success has attended its labours, and we yet hope to see artificial propagation more appreciated, and more extensively carried out in our own country; but we are afraid this will not take place until there is a still greater falling off in the returns of fish from our Scottish rivers, when a greatly diminished rental will make our salmon proprietors converts to artificial propagation.

THE SMOULT OR YOUNG SALMON AFTER MIGRATION.

In descending the river to the sea, smoults congregate in shoals, rise greedily to the fly, and become an easy prey to the angler. They proceed slowly, unless a heavy spate occurs at the time, which soon sweeps them into the ocean. The marked pond-bred smoults have been traced all the way from that place to the salt water—a distance of at least twenty miles. Even those marked by the rings have been brought ashore in the fishermen's nets many miles below the pond, and others have been observed, by the men in charge of the still-net at Broughty Ferry, having the ring attached, ten or twelve days after marking. What becomes of the fish after it reaches the sea is as yet matter of conjecture. All we know for certain is, that after six or eight weeks a number return to the same river, having increased in weight, from half-an-ounce to between two and five pounds. What they got to feed upon while in the salt water, naturalists are not agreed upon,* but there is no doubt that in the sea their food is abundant. Not a few naturalists have said that salmon travel a

* See Appendix.

long distance after reaching the salt water ere they return, but this we think cannot be the fact, unless they were to swim at a great rate, which would be against their acquiring the weight and size which they attain to in so short a time. Those fish, however, that do not return until next year may be greater rovers; but we are almost certain that this cannot be the case with the early grilse—they must find their food not far from the river's mouth. We have already stated the particulars of the smoults returning as grilse.

RETURN OF SPAWNED FISH OR KELTS AS CLEAN SALMON.

His Grace the Duke of Athole, for a number of years past, has been experimenting on these fish, and has succeeded in establishing the fact that some of them at least return to the same place in five or six months, having gained in weight, in that time, from 7 to 10 lbs. These kelts are caught with the rod, and marked with a gutta percha or copper ticket, which is fastened round the tail of the fish. This ticket has the Duke's name and a number on it. When the kelts are caught they are weighed, a ticket is attached, the fish is then returned to the river, and the number, weight, and time of capture are entered in a journal kept by his Grace.

When the fish are again caught ascending the river as clean salmon, the fishermen send the tickets to the inspector of the river, who forwards them to his Grace. We shall here notice a few of those captured, with the weight and date when taken as kelts or spawned fish, and the date and weight as clean fish:—

1859.   CAUGHT AS KELTS OR SPAWNED FISH.
No. 21—Feb. 14. Caught a little below Logierait,.......10 lbs.
No. 76—Mar. 2. Caught a little above Dunkeld,......11½ lbs.
No. 95—Mar. 29.   Do.   do.   do.   ......12½ lbs.

1859.   CAUGHT AS CLEAN SALMON.
No. 21—Aug. 18. Caught at Channelbank,...............17 lbs.
No. 76—Aug. 18.   Do.   do.   ....17 lbs.
No. 95—Aug. 12. Caught at Elcho Castle Fishings,* 19 lbs.

Many more spawned fish have been marked by the Duke, and from statements furnished us, some of those have returned from the sea, and been captured in the river as clean fish—having remained a much shorter time in the sea than those in our list; but as we have no correct dates

* These two stations are situated on the Tay, a few miles below Perth.

of the time of their capture, etc., we are obliged to reject them. It is clear, however, from what we have stated, that at the end of four or five months from the time of their capture by the rod as spawned fish, a number return from the sea to the river as clean fish, having increased 7 lbs. in weight.

DO SALMON FALL OFF IN CONDITION ON ENTERING THE FRESH WATER?

That all fish of the salmon kind that are migratory get into worse condition after being some time in fresh water, no one can doubt who is at all acquainted with the habits of the fish; but it is not so much the change of water and food as the state that the fish is in at the time when it seeks the fresh water. We have already shown on the evidence of Mr Ffennell, and from our own experience, that the winter or closetime running fish are in prime condition for the table, if not for the market, after being many months in the fresh water. The fact is, the good edible condition of the fish depends entirely upon the state of the development, and size of the milt or roe in the fish. When fish are kept from entering the river by a want of a fresh in the water to enable them to leave the sea, even there the development of the spawn takes place the same as in the fresh water, and they assume the red colour which they have while on the redds. At that time, they are not fit for food, but, notwithstanding, are eaten as kipper. On referring to Mr Johnstone of Montrose's examination,* it will be seen that it was this full development and growth of the milt and roe that led him into the absurdity of believing that salmon spawned in the sea. The clean fish that ascend the Tay in the months of November, December, and January, and are caught in Loch Tay and other feeders, in spring, at the commencement of the season, are good and excellent for the table, albeit a little brown in the colour, which may injure them as a marketable fish; but the spawn in these fish, although developed, has not become as yet of any size, and of course the fish has not fallen off much in condition.

KELTS, OR SPAWNED FISH.

It is well known to anglers and fishermen that kelts do not immediately after spawning return to the sea. The fish become very weak after the operation; and they afterwards drop down

* See Appendix.

from the redds to the first deep and quiet part of the river they can find, where they remain until they have recovered some strength. They then continue dropping down the river, descending from pool to pool, avoiding the rapid running places as much as possible. At this time they feed greedily, and are a great annoyance to anglers, during the spring months, as they rise at any sort of fly; and, although not so difficult to land as a strong clean fish, they sometimes take more time to land than the angler is willing to spare. Being at this time such voracious feeders, there is no doubt but, like Saturn, they devour their own offspring, for a parr is a good bait for them. This was one of the excuses of the fishermen for not returning them to the river after being taken in their nets, for, within a few years back, almost all the kelts caught were not returned to the river, and of course large salmon were becoming scarce. Matters now, we are glad to say, are changed, as far as the Tay is concerned, since the passing of the last Act of Parliament, and the result is that the large fish are getting more plentiful. There is no doubt that kelts or spawned fish improve much in condition in the fresh water before they reach the sea. In proof of this, we will instance a case which occurred to ourselves last year:—On the 4th of May, 1861, we hooked a fish on the Stanley water, and, as the fishermen assured us the kelts had all left the river, we were flattering ourselves with the prospect of landing a fresh run salmon. He fought well for liberty, and twice did he make his appearance, and was pronounced clean by the fisherman who had the gaff extended, when his eye caught the glittering copper medal attached to the tail. We ordered the gaff to be laid aside, but had to give the fish another run before it could be lifted out without injury. It turned out to be a male fish, and had a copper medal fastened round his tail, on which was stamped Athole, No. 78. He weighed 16 lbs.; and but from being a little lank at the tail—which might have been caused by the copper wire, which had cut the fish nearly into the bone,—he might have passed muster for a clean fish (there were no maggots on his gills), as the fishermen thought he was, until we showed them the vent, which was too much enlarged for a fish that had not lately spawned. After having carefully returned him to the river, we noted the number, and apprised the Duke of the circumstance, when his Grace was so obliging as to send us the following account of the fish:—

"DUNKELD, *May* 9, 1861.

"The Duke of Athole presents his compliments to Mr Brown, and is very much obliged for his civility in forwarding particulars of the capture of a marked fish. The Duke thinks it may interest Mr Brown to learn that the kipper No. 78 was caught on the 1st of April by Mr Evans above Logierait. It then weighed 13½ lbs."

We see that salmon do, in certain states, improve rapidly in condition in the fresh water. Every one has heard of a well-mended kelt, and not a few of them in the spring are forwarded to the large towns and sold as clean fish, but generally at a lower figure than the clean salmon. They are bought by thrifty housekeepers who must have a salmon dinner like their neighbours at the opening of the fishing, but who are not inclined to give the big price. A well-mended kelt is as clean and silvery-looking as a clean fish, but the head looks large, as the body is not filled up, and upon opening the gill covers, a white-looking maggot may be seen firmly adhering to them. The fish when cooked emits a rashy smell, and as an article of food is unwholesome.

No one on any pretence should be allowed to take spawned fish from the river, but when caught should return them as little injured as possible; and anglers who use the gaff in their capture should be informed upon. Sometimes, however, the hook will wound the gills of the fish, which in every case proves fatal; but these instances are rare, and to enjoy spring fishing without hooking kelts is impossible. The folly of not returning them to the river must be evident from the speedy return of at least a number of these fish during the same season, so greatly increased in weight and condition, as has been proved by the experiments of the Duke of Athole.

The efficient protection of our salmon rivers, and a free passage up the stream to the spawning beds cannot be too strongly insisted upon. The present law on this subject in Scotland is insufficient for this purpose, or it is not acted upon. We would instance the river Ericht at Blairgowrie, which was formerly one of the most famous salmon rivers in the country, but at the present time a salmon is seldom seen in its waters, and all this is owing to mill-dams which have been erected upon it; some of these are of a height that, even at the greatest floods, salmon find it impossible to surmount. All that is wanted at the dam-dykes is salmon ladders; but who is to be at the expense of these erections? is the question; and until a bill, similar to the English Fisheries Bill, be extended to Scotland, no remedy will be found for this beautiful little stream, whose waters are also poisoned by chemical mixtures which are discharged into it from manufactories on its banks. By obstructing the free passage of the salmon to the spawning grounds, by "the mill-dam, the torch, and the spear," as Colonel Whyte has observed, "the rivers of Halifax, that twenty years ago were teeming with salmon, are now utterly destroyed; and also the rivers of Canada, which at no distant date were swarming with sea-trout and salmon, have fallen so much off, from the same cause, that Government has at length interfered to stay the destruction; and, in not a few instances, recourse has been had to artificial propagation to re-stock some of the rivers." Some theorists affirm that salmon and civilisation cannot exist together, as draining and limeing will, in the long run, destroy the fish in any river along whose banks it is carried on; but this cannot be the cause of the falling off of the fish in the rivers which empty themselves into the St Lawrence, as the agriculturist has, as yet, scarcely broken ground on its banks. When the redskin possessed the country, although he caught fish at all seasons when he could get them, and by every method known to himself, he sought them only to satisfy the cravings of nature; but when the whiteskin made his appearance on these waters, all the methods known to civilised man were called into requisition to capture the fish as a marketable article; and as there was no law to tell him how or when he was to take salmon, the destruction must have been great. The saw-mills arose, also, with their insurmountable dam-dykes, and prevented the ascent of the geese that were willing to lay more golden eggs; and in less time than a quarter of a century, these rivers that were swarming with fish had fallen so low that Government was at last aroused for their protection, since which time salmon have commenced again to multiply—hence it is clear that, without protective laws, the genus salmo would soon be extirpated in any populous country. No rivers have suffered for want of protection more than the English rivers, but there is little doubt that not a few of them, if the late Act of Parliament be sufficiently carried out, will ere long be again well stocked, although it is our opinion that others will require the assistance of artificial propagation to insure that end.

ATTEMPT TO RAISE THE SALMON FROM THE OVA TO THE GRILSE STATE ARTIFICIALLY.

When none of the smoults which were marked with the silver rings had been caught returning as grilse, we resolved to try if it was possible to rear the salmon from the ova up to the grilse state, under our own eye, by means of a salt-water pond into which the sea ebbed and flowed. We found a pond well fitted for the purpose, which had been made at great expense at Stonehaven, in Kincardineshire. This establishment was originally intended for the purpose of hatching and rearing the fish from the ova up to the full-grown salmon. It consisted of two ponds—a fresh-water one, 40 × 20 yards, and a salt water one, 60 × 30 yards. The water was taken by a pipe from the Water of Cowie, which flows past the place. Where the pipe conveying the water ended, there was a ditch cut, into which the water flowed, and supplied the fresh water pond. Between the fresh and salt water ponds there was another cut for conveying the fresh water into the salt. This last cut had a fall of eight feet. The distance between the ponds is about thirty yards. A strong iron pipe, eighteen inches in diameter, is carried from the salt water pond to the sea—the mouth of the pipe is placed in the sea, as far as half-water mark, by which means the water is frequently running in or out. The pond, when the water is at the lowest, is about three feet deep, and is well built with projecting stones along the sides, for shelter to the fish. At high water it has six or eight feet added to its depth. The water at the place where it enters from the sea is as salt as the ocean, but at the sides it is brackish, as a considerable quantity of fresh water filters into it. Some years ago, when this pond was completed, a few pairs of spawning fish were turned into the cut which empties itself into the fresh water pond, and we believe they did spawn, and dropped down into the sea pond; but they were not allowed to remain long there, the temptation was too great for the poaching propensities of a set of prowlers that are to be found in all places. We learned from persons on the spot, that the ova hatched, and disappeared likewise. Multitudes of parrs about two inches long were seen, but as there was nothing to prevent them from descending the stream into the salt water, no doubt the most of them did so and met their fate. The experiment wanted much to insure success, and of course proved a failure.

In the spring of 1860 we went and examined these ponds, and having anew covered the mouth of the sea pipe with galvanised wire grating sufficiently small to retain a smoult, we selected five smoults which were reared at Stormontfield from salmon ova. One of them was four years old—it was three years old before it became a smoult, as mentioned previously, and had only grown, although well fed with boiled liver, to eight inches in length. The four others were two year old smoults from five to six inches in length. With a considerable deal of trouble, by having seven changes of water ready at the stations—the distance to be carried by rail being about sixty miles—we got the fish safely put into the salt water pond in excellent health. When put in they swam slowly for a few minutes amongst the brackish water at the side, and shortly afterwards they darted into the pure sea water. The pond at this time was swarming with very small fry of sea fish, (we did not discover what kind of fish they were the fry of,) which came in through the grating with every tide, and we saw abundance of food for the young salmon. We left them under the charge of Mr Gray, the tacksman of the salmon fishings at the mouth of the Cowie, and also the men employed by him, who lived on the spot, and they promised to see that no harm came over them. The fish were put into the pond on the 30th of June, and on the 8th of July our correspondent writes: "I have the great pleasure to inform you that your fish have been seen in the evening jumping at the flies. I have no doutbt but they are delighted with their new quarters as they have everything that they can desire. The salmon fishers seem to take a great interest in them."

July 28th.—The fishermen reported that they had seen the whole of the fish at the mouth of the pipe—their favourite resort. The largest appeared to be about the size of a herring.

August 5th.—The four year old smoult was reported to have been seen by one of the fishermen, and that it had the appearance of a sea trout of $1\frac{1}{2}$ lbs. weight, but the rest of the fishermen said it was not so heavy.

On the 18th of August we went north, and were so fortunate as to see three of the young fish; they had grown double the size. We looked in vain for the large one, but owing to the ripple on the water it was difficult to see any distance from the side. The winter of 1860-61 turned out very severe, and our correspondent's communications were always "fish cannot be seen, owing to the stormy weather." On the 15th of April, our correspondent writes — "that I had the good fortune, yesterday morning, to discover one of your lost family, it was lying at the mouth of the pipe, with its head to the stream. I had as good a view of it as I could wish. It appeared to be less than the large one that I saw lying in the same place last summer; but I may state that it would be about double the size and weight of those you saw when here in August last;" so that this fish had increased to the size of a sea-trout of 1 lb. weight in about ten months. Our next visit was in August, when we were told by the fishermen that they were afraid our fry were all gone, as a notorious poacher had been seen early one morning whipping the pond with his flies, and none of the fish had been observed since. We felt very vexed at the news; but from the unprotected state of the place, and the great temptation there was to persons who live by poaching to make a few shillings by the sale of the fish, we were not at all astonished. The fish had grown marketable, and that they were in the pond was well known to many people in Stonehaven, and this circumstance could not escape the notice of these lawless characters. Had we only been able to get one of them to preserve, so as to show the history of the fish under artificial culture—from the ova to the grilse—our labour would have been repaid; but as it turned out, the experiment was a failure, as the fruits of it could not be made patent. We learn, however, from it, a fact well known already—that the smoults of the salmon grow much more rapidly in the salt than in the fresh water, and that this is chiefly owing to the abundance and variety of food contained in the former, and, further, that it is possible to rear the fish up to the grilse state artificially. The little river of Cowie, upon which these ponds are situated, belongs to the heirs of the late Alexander Baird, Esq. It was at one time swarming with salmon, but since 1825 there has been a stake-net at its mouth; besides, on looking seaward, the black bladders of bag-nets are to be seen bobbing in every direction, which renders it impossible for any fish during open time to ascend the stream.

Stake and bag-nets, or fixed engines of any kind, for the capture of salmon in the sea—dam-dykes, without openings or stairs in them—saw-mills, the owners of which are allowed to send their dust into the rivers—bleachfields and dyeworks, the most of whose proprietors make no scruple to pour their poisonous chemicals into the pure stream direct, so that not a smoult can attempt to pass without being destroyed; these are chiefly the causes of the great falling off of the salmon in our rivers. Want of sufficient protection to the spawning fish is another cause, without doubt, but this neglect is nothing in comparison to the other evils; the poacher may do much harm, but the first-named evils are the sure methods of destroying the breed altogether.

APPENDIX.

EXTRACTS FROM THE REPORT BY THE SELECT COMMITTEE OF THE HOUSE OF LORDS ON THE SALMON FISHERIES, 1860.

First.—As to the law affecting salmon fisheries in Scotland.

By the law of Scotland the right of salmon fishing is *inter regalia*, and is vested in the sovereign, *jure coronæ*. The salmon is not, however—like the whale or sturgeon—a royal fish, so as to give the Crown an actual right of property in every salmon that is caught; but the right of the Crown is to the salmon fishings in Scotland, which, as observed in a case that lately came before the House of Lords in its judicial capacity, "appears to be a common and well understood description of the subject of claim."

Salmon fishings have from a very early period been the subject of grants by the Crown, and the fishings in rivers which yet remain vested in the sovereign are probably not of great extent.

The grants by the Crown of salmon fishings in the sea are not so numerous as those of fishings in rivers; and although some grants have been made of fishings in the sea, there appears to be altogether a considerable extent of coast upon which the salmon fishings still belong to the Crown.

The nature of the title of the Crown to salmon fishings in Scotland being that which has been above described,* and being vested in the Crown—not merely as trustee for the public, as has been contended, but as part of its patrimonial estates—it is next to be considered how the right to any particular salmon fishing may be acquired by a subject. Such a right may clearly be obtained by an express grant from the Crown; and it has been stated to the Committee by the Lord Advocate that if there be a title from the Crown *cum piscationibus*, followed by a possession of salmon fishings by the use of a net and coble, cruive, yair, or other mechanical contrivance, for forty years, a presumptive right to salmon fishings would be acquired, but that rod or spear fishing will not be sufficient to found a presumptive title. It has been proved in evidence that many grants of salmon fishings in the sea, and in estuaries in Scotland, were made by the Crown before the transfer of the management of the Crown property to the Commissioners of Woods, and the list of such grants which was produced to the Committee probably does not contain references to nearly all the grants which have been made, or titles which could be maintained against the Crown.

The statutory restrictions on the salmon fishings in Scotland, as regards the use of fixed engines, may be considered, first, with reference to rivers above the flow of the tide—that is to say, from the source down to the place where the flow and ebb of the tide is perceptible; secondly, with respect to that portion of rivers which is within the ebb and flow of the tide; and, lastly, with reference to the sea coast.

The general statutes by which these restrictions were imposed are very numerous—they commence with an Act

* See Blue Book.

of Robert I., 1318, chap. 12. The Committee gather that the effect of these statutes, as regards the use of fixed engines, is shortly as follows:—

1st, Cruives are legal from the source of a river down to the point where the ebb and flow of the tide begins, provided the person using such an engine has an express grant from the Crown of the privilege of fishing in that manner, has exercised that privilege, and observes the regulations of the statutes as to cruives, the principal of which are that they shall be kept open from Saturday till Monday, and that the hecks or bars of the cruive boxes shall not be less than three inches apart.

2d, Cruives, yairs, and all other fixed engines, are illegal in a river from the point where the flow and ebb of the tide begins, down to the sea.

3d, Fixed nets are not illegal on the sea-coast, and a person having a grant of salmon fishings on the sea-coast cannot be interdicted under the statutes by another proprietor of salmon fishings, or be prevented by the Crown from using such nets.

The Tweed and Solway are exempted from these Acts.

The question as to the extreme limit of a river towards the sea, within which the prohibitions against fixed engines apply, is one which must be determined by a jury in each case as it arises, and a variety of circumstances would have to be taken into consideration, varying with the locality, so that the result in one case would be but little guide towards a sound conclusion in another.

With respect to the annual close-time, or period within which it is allowed to take salmon. By an Act of the Scottish Parliament, passed in 1424, it was forbidden that any salmon be slain from the Feast of the Assumption of our Lady until the Feast of St Andrew in winter. The dates of these feast days being corrected according to the new style, the close-time established by the Act of 1424 was from the 15th August until the 30th November (N. S.)

But by the Act 9 Geo. IV., c. 39, 1828, commonly called "The Home Drummond Act," the annual close-time was altered, so as to last from the 14th September until the 1st of February. This Act does not apply to the Solway, or its tributaries, or to the Tweed, and the close-time in the Tay was, by the Tay Fisheries Act of 1858, enlarged, so as to commence on the 26th of August and to end on the 1st January, (this is a mistake, it should have been the 31st of January.) With these exceptions, the annual close-time for salmon fishing in Scotland is still regulated by the Act of 1828.

The Committee, after noticing the evidence generally, state, "After giving their best consideration to the evidence submitted to them, the Committee are of opinion that, with a view to the improvement of the salmon fisheries in Scotland, all cruives and fixed engines, of whatever kind, both in rivers and in the sea, should be abolished, and, at all events, no new fixed engines, of any description, should be permitted to be erected," etc.

James I., 1424, c. 35, (Repealed by George IV., c. 39):—"It is ordained by the Parliament, and forbidden by the King, that onie salmound be slaine frae the Feast of the Assumption of our Ladie quhill the Feast of St Andrew in winter, nouther with nettes, nor cruves, nor nane otherwaies, under the paine put upon slayers of red fish, quilk alswa the Justice-Clerk sall gar inquire."

There are other three Acts of James I. regarding the salmon; one of James II.; two of James III.; three of James IV.; one of James V.; one of Queen Mary; three of James VI.; one of James VII.; two of William III.;

and one of Queen Anne. None of these Acts interfered with the annual close-time of James I.

Very great care was shown by our ancestors in the protection not only of the salmon but of the fry, for we find in the Act of James VII., c. 24, May 30th, 1685— "Item: That all millers that slay smoults or trouts with creels or any other engine, or any who dams or laves, shall be punishable as slayers of red fish, conform to the (37) Act Parliament, 5, King James III.; and where the transgressors has no means, they are appointed to be put in prison, irons, or stocks, for the space of one moneth, upon their own expenses; and if they have it not of their own, to be fed on bread and water, conform to the 89th Act of Parliament, 6, King James VI.," etc. The steeping of lint in rivers, lochs, or burns, where fishes are, is also forbidden by the same Act.

EXTRACTS FROM MINUTES OF EVIDENCE.

WM. JOSHUA FFENNELL, ESQ., ONE OF THE COMMISSIONERS OF FISHERIES IN IRELAND, ON WHAT ARE CALLED BARREN FISH.

With regard to salmon that enter the river in December and January, Mr Ffennell (answer 2538) says— They are fish in which the ova is fully developed, but not grown large. (2540)—They would spawn in twelve months. (2541)—He also says that December and January are the principal spawning months in Scotland.

2551—I apprehend that all the monster fish are old fish.

2555—LORD LOVAT: Do the early spawned fish return to the sea and come again? Yes, sooner of course than the late fish. With respect to the spawning operations, it appears at first strange that some fish should seek to make the fresh water their habitation for the whole of the summer, while other fish remain in the estuaries and the sea until within a few weeks of spawning, and come out of the sea quite ready to spawn. It presents itself to my mind as one of the wonderful arrangements of nature—that it is for the purpose of causing the fish to distribute themselves through the whole of the waters. That is their tendency, for we have the fish in summer from the spring going up the rivers, and killing themselves to get over obstacles; actually jumping at weirs and locks and all sorts of things, while a number of fish remain quiet in the sea, and do not come into the fresh water at all till they are just ready to spawn. We find also that the earliest spawning is almost at the top of the water. I think it is as clear as possible that the object of the law that governs them is, to cause the fish to distribute themselves throughout the whole length and breadth of the water. Some people have thought it extraordinary that there should be these December and January fish—we have it before our eyes. I have often taken them myself in the lakes in the following summer in all stages of ova in progress of development. We have them under our eyes in rivers. The fish come up in February and March, and we know that they do not spawn till November and December, following, after all, very much the same time before spawning as the December fish that spawn in October. It is about the same time.

2556—LORD POLWARTH: You think that the fish enter the river in December and remain in the river till October, and then spawn? Yes, no doubt of it. They are the most valuable spawners.

2557—Viscount Hutchinson: It has been stated to the Committee, as the opinion of one of the witnesses who have been examined, that those earlier fish in the Scottish rivers are barren fish. You do not coincide in that opinion? I can answer for it in Ireland, that they are not barren fish. I have tried many experiments about it. Before we allowed January fishing in the Carra, my colleague, Mr Barry, and I went down, and we caused the cruives to be put down in the river Carra, which is one of the earliest rivers. We assembled the Board of Conservators, the fish brokers, and all the people interested in the river. We had the cruives put down at night on the 23d of November, and in the morning we had seventeen salmon. Fifteen of those salmon were as beautiful marketable fish as could be. There were only two bagots among them. We opened every single fish. We had a *post mortem* examination on the spot before all these people, and every one of those that were female fish (they were about in equal proportion) had the ova quite developed. Of course it was very small, smaller than mustard seed. The lobe was perfectly formed, and the little seed in it. The milt of the males was in a corresponding state. I have tried them upon many occasions, and there is no doubt about it, that they are not barren.

2558—Lord Polwarth: That spawn would have ripened and developed in the fresh water? Quite so.

2560—Earl Ducie: You say those fish you caught in November had imperfectly developed ova in them? Perfectly developed; it was very small.

2561—Those were the ova which would have been deposited in the succeeding October? Yes; I have seen the salmon taken, and I have caught them myself in the early lakes quite a copper colour; the ova not very far advanced; not a good-looking fish to send to market, but a right good fish to eat.

Mr Ffennell on Stocking a River.

2604—CHAIRMAN: A comparatively small number of fish, if allowed to ascend the river and spawn in security, would be sufficient amply to stock the river, would it not? I think not. I think it requires a good stock of fish amply to stock a river, though people have spoken of the many thousands of ova (something like from 11,000 to 17,000) in a large salmon; but from the very commencement there is a great waste, and I believe one reason why they are provided with such a quantity is to meet this waste. In the first place, there is a great quantity of ova not impregnated—that I have proved myself, by taking it up out of the gravel. The greater portion of it was quite opaque, and the milt of the male had not come in contact with it at the right time. There is another large proportion of it that the fish fails to cover in the gravel. It is carried away by the stream either into deep water or into muddy places and is lost. Unquestionably, when the little animals come to life, that red bag that is attached to them for three, or four, or five weeks is a great attraction, and the smallest trout eat a quantity of them. All fish live upon each other. Ducks, particularly wild ducks—which abound in Ireland—destroy a vast quantity of ova. Every spawning bed in the winter is a favourite place to go and watch for wild ducks. You see them with their head under water, with their heels up, rooting the ova out of the spawning beds. Wild ducks feed immensely upon it. Then, when the fry are going down, a great many are devoured, and also when they get to sea. I once saw twenty-six salmon fry taken out of what we call a "black pollock" in Ireland, which is called the coal fish—a big fellow weighing 15 or 20 lbs., that was killed in the Bay of Killala, outside of Ballina, when the fry were going down: to meet all these things the fish are given a greater multitude of ova. I do not think that a few fish will stock a river. I am quite satisfied in my mind that it will not. I do not know a river in Ireland that is half-stocked with salmon fully to develop its resources.

2612—LORD LOVAT: Is spawn hurt by frost? If it is exposed. I do not think that, unless it is quite bare, it is hurt by frost.

2624—LORD LOVAT: Could you take salmon up 30 or 35 feet by means of a ladder? You could, or any height.

SALMON SPAWNING IN THE SEA—MR JOSEPH JOHNSTONE, MONTROSE.

3439—LORD COLVILLE: Do you think that salmon spawn in the sea? My faith in that has been very much shaken. I have some doubts that they do spawn in the sea.

3440—You once thought so? I have some doubts now. I did not believe at one time. I begin to think they do.

3441—LORD STANLEY: Why do you think so? For instance, I am tacksman of a small fishing near the Tweed, that they call Marshall Meadows, within the limits of the Tweed Act. It is on the march between Scotland and England, and after the Act was passed last year, I was very particular at the end of the year to ascertain what fish there were. I had never seen them. At the last day of the fishing I got a great many brown salmon quite full of spawn.

3442—Do you mean kelts? No; the fish full of spawn, that you would think were ready to spawn; and I ordered them to send them on to Montrose, that I might see what they were like. I got some twenty or thirty which were very full of spawn, very brown, and they were fish close to spawning.

3443—EARL INNES: Are you aware that the Tweed was very dry till the close of the fishing season, and that the fish could not get into the river for want of water? I do not know.

3444—Lord Stanley: Do you often take brown fish in the sea? I never saw them but there.

3445—Had they the same colour that fish have that have been a long time in the river? The very same as if they had been lying in the river for two months.

3446—Earl Innes: On the part of the coast of which you speak is there not a great quantity of sea weed? Yes.

3447—Do you not think that the colour of those fish was attributable to their lying among the sea weed? I think it has that tendency; at the same time the fish were so very full of spawn.

3448—Was that at the end of September? It was just the last two or three days of fishing.

3449—Then, of course, the fish would be full of roe, it is natural to suppose? I never saw them so full in the Tweed before.

3450—Lord Stanley: Do not you suppose they were fish waiting for sufficient water to ascend the river? I do not know, or whether they were going to spawn in the sea, that they were in that condition. There is a strong current and a small grave outside in the sea.

3451—Is the reason which induces you to hesitate in the opinion which you formerly came to, that salmon do not breed in the sea, because at the end of the season last year you took those brown fish which were full of spawn, and therefore you think they spawn in the sea? I would not say positively, but I have a certain amount of belief in it.

3481—Lord Lovat: Do you consider that the fish caught in Lunan Bay are Esk fish or Tay fish? I believe they are from both rivers. I believe there is plenty taken that never was in any river.

Mr John Hector, Tenant of Salmon Fishings on the Sea, *ex adverso* certain Lands in Kincardine—Commenced Salmon Fishing in 1819 or 1820.

2754—Lord Colville of Culross: Do you think any salmon spawn in the sea? I think they do.

2755—Did you ever find one spawning in the sea? No. Judging by the quantity of fish that I have seen in the sea, I should think they could not all go in to the rivers.

2756—Chairman: Have you ever found any smoult or salmon fry in the sea? I have.

2757—Lord Colville: Have you ever found a fish very near spawning in the sea? I have; near the end of the season. . . . .

2758—Do you think that those fish that are very full of roe would spawn in the sea if they could not get to the rivers? I cannot say; I have not tested it. My opinion is, that if all the fish came from the sea into the rivers, for the purpose of spawning, the rivers could not contain them. The sea is teeming with fish; I have seen great numbers in it.

2759—Do not you think those fish that are full of roe are prevented from going into the rivers, and that is why they are so found full of roe in the sea, that they cannot find their way into the rivers, or are prevented by stake-nets? They are not prepared for spawning till the end of the season. My opinion is, that the fish taken in the summer season are not on their way to the rivers for the purpose of spawning; that is my candid opinion; and we kill a great deal of fish in the sea. I have no hesitation in saying, from my experience in fishing for 40 years, that they would not go to the rivers. . . . .

2760—Chairman: Does it consist with your knowledge that experiments have been made with a view to the propagation of salmon in salt water, and that those experiments have failed, and the ova never have become productive? Both in salt water and fresh water—I am afraid it will not succeed; that is my opinion.

2761—I asked you whether you were aware that experiments had been tried to propagate salmon in salt water, and that those experiments have not succeeded, inasmuch as the ova had not produced the young salmon? They have not succeeded.

2762—Are not you aware that those experiments have succeeded frequently in fresh water? In bringing the fish to life they have succeeded.

2763—Would not that tend to show that it was impossible that salmon should propagate in salt water? There may be different qualities of fish. There may be sea fish and there may be fresh water fish.

Food of the Salmon in the Sea—Professor John Queckett, College of Surgeons.

3641—Earl Ducie: Is it your belief that a salmon which has been bred in a river travels to any distance from it when he enters the sea? I think salmon travel along the coast, but they endeavour to return to the river where they were bred.

3642—Chairman: They usually do so? Yes, as far as I can make out. They travel some considerable distance into the deep water, and into the sea, for their food, which is, I believe, essentially the ova of the echinus or sea-urchin. That is what I have always traced in their stomachs, and I know, from the locality in which those sea-urchins are found, that the salmon must go into deep water, some considerable distance from the shore, to get them.

3643—At what depth does the echinus, upon the ova of which the salmon feed, live in the sea? They live in from six to twenty fathoms.

Professor T. H. Huxley, at the Government School of Mines.

3733—Earl Ducie: Do you know what the habits of the salmon are in the sea—upon what he lives? There is much reason to believe that those animals live chiefly upon entomostracous crustaceæ, which are found in the sea in very great abundance. I believe that is one reason of the great increase of size which the salmon attain in the sea; because in the rivers the only food of this kind they can obtain consists of insects and insects' larvæ, which are comparatively few; but directly they come to the sea there are great patches of water where crustaceæ are found in abundance.

3734—In fact, they swim in a species of animal soup? Yes.

3735—Earl Cawdor: Are those crustaceæ found in deep water? Yes, it would be quite possible for the salmon to obtain such food there.

3736—Earl Ducie: What depth are the crustaceæ in the water? They are at the surface in deep water. Very often you find the surface completely thick with them.

3737—One of the witnesses has stated it as his belief that the principal food of the salmon is the ova of the echinidæ? I doubt that very much. I do not know how the salmon are to get them to begin with. If you examine the surface of the sea, you find the embryos of echinodermus mixed up with other things, and there is a far greater multitude of minute crustaceæ and molluscs. The salmon can only open his mouth and take what comes into it; he cannot separate them. You do not get the ova or embryos of the echinidæ in distinct patches by themselves.

3738—Those ova are not deposited at the bottom of the sea? No; the ova are very rapidly hatched, and as soon as they are laid they become fry or swimming embryos.

3739—Are they ovoviviparous? Hardly that; but still the eggs are hatched very rapidly.

3740—Do the salmon feed upon the embryo? I should imagine they must do so to a certain extent. A salmon can only feed upon these small creatures by opening his mouth and taking what comes into it. Those who have examined the stomachs of salmon fresh from the sea, affirm that they contain great multitudes of minute crustaceæ.

3741—There is nothing in the nature of the food of salmon to confine them to the immediate vicinity of the shore? I should say not, so far as I am aware.

3742—EARL CAWDOR: Do these crustaccæ deposit their ova at the bottom of the sea? It is those crustaceæ themselves which are the prey of the salmon, and not the ova. They are very minute animals, and very abundant; and the most of them carry their eggs about with them.

3743—Are the ova of the echinidæ deposited upon the bottom of the sea? They may be so deposited for a short time; we have no positive evidence that salmon feed particularly upon them, but we know that they feed upon these entomostracæ or small crustaceæ.

3744—LORD LOVAT: Is a foul salmon long in getting clean after he gets into the sea? I am not aware; I think it would be very difficult to ascertain that point in any way.*

* See the Duke of Athole's Experiment.

GLASGOW: PRINTED BY THOMAS MURRAY AND SON.

www.ingramcontent.com/pod-product-compliance
Lightning Source LLC
Chambersburg PA
CBHW031618170426
43195CB00037B/1099